EXPEDIENT B&E

Carl Hammer

EXPEDIENT B&E

Tactics and Techniques for Bypassing Alarms and Defeating Locks

PALADIN PRESS
BOULDER, COLORADO

Expedient B&E:
Tactics and Techniques for Bypassing Alarms and Defeating Locks
by Carl Hammer

ISBN 0-87364-688-6
Printed in the United States of America

Published by Paladin Press, a division of
Paladin Enterprises, Inc., P.O. Box 1307,
Boulder, Colorado 80306, USA.
(303) 443-7250

Direct inquiries and/or orders to the above address.

Contents

Note

As this book was originally written in Europe, all measurements follow the international metric system. No measurements are absolute, however, as they always must be adapted to the measurements of the actual lock or device you are working on.

1 millimeter = 0.039 in.
1 centimeter = 0.39 in.
1 meter = 39.37 in.
10 meters = 32.81 ft.

Introduction

Breaking and entering is a fairly common intelligence technique employed in all types of secret warfare. There have been many successful entry operations that have been of vital importance. One such operation, now well documented, took place in Chicago in 1941.

A team of operatives led by naval intelligence agent Willis George performed a covert entry into the office of Stephen Ziggly, a German spy who posed as a businessman. The search of his office produced sufficient evidence to break up one German spy ring and put Ziggly in jail. (After the war, Willis George wrote a book about his experiences as a government burglar. This book has now been reprinted by Paladin Press under the original title *Surreptitious Entry*.)

On the other hand, a failed or, even worse, detected and exposed entry operation could easily be a public humiliation for the intelligence service employing the exposed field operatives. This has happened on several occasions, always with serious consequences.

For these reasons, every entry operation must be prepared and executed efficiently. Although the situation and the target area always will be different from one operation to the next, this volume aims to point out the most commonly used entry techniques and explain how to employ them in the most expedient manner, whatever the situation. The field operative must himself consider which options would be valid and which ones would be unsuitable.

An entry operation necessarily demands that the operatives somehow manage to enter the premises of the target. This is usually not easy, as there will be locking devices and possibly also alarm devices to prevent intruders from entering or sound when an intruder has been detected. The first priority is therefore usually to bypass the alarm system, and then to defeat the locking devices. The latter will be accomplished either by picking locks or breaking through them.

This book does not claim to teach everything about lock picking. That is an extremely specialized subject, demanding a complete knowledge of the mechanism of all locking devices as well as a year or more of constant practice. Consequently, lock picking is best left to the experts, who devote all their time to it. This book does, however, give an overview of the techniques available to the operative who is called upon to perform an entry operation, including details on how to pick most common types of locks. This book also details many not so well-known techniques for clandestinely entering a house. Once a resourceful operative has studied this information, the existence of an alarm system or security system will provide no guarantees of keeping him out.

Lock picking will be fully described in Chapters 2 and 3, while other entry methods are described in Chapters 4 and 5. Most of the latter half of the book is devoted to alarm systems and alarm sensors. The text explains how alarm systems can be circumvented (i.e., defeated) by avoiding all sensors and switches and how they can be defeated by other means, such as frustrating them. These two methods, circumvention and counteraction, are detailed for most types of current alarm systems, as well as a few upcoming types, which we will probably see more of in the future.

It is remarkable that so many present-day alarm systems, including numerous American do-it-yourself security systems, are characterized by an unfathomable naivété.

Although the electronics might be highly advanced and truly state-of-the-art, the actual application of these systems is generally easy to defeat. This is most likely the result of the extremely large number of companies wishing to cash in on the security market, each one trying to attract attention with the latest gimmick, whether it works under actual field conditions or not. In the future, other access control systems might turn out to be more common than the leading ones today. Devices for identifying people by fingerprints and handprints, as well as voice identification systems, are currently being developed. Some of these systems are already in limited use. Despite this, it is quite certain that the present alarm and locking devices will remain in use for a very long time indeed. After all, the present locking technology for the most part relies on and dates back to the nineteenth century.

It is likely that there will be no clear-cut borderline between alarm devices and locking devices in the future, as these two groups are approaching each other a little more each day. This can be seen clearly in the wide variety of electronic locks and card locks available today. I expect that we will see more of this integration in the future. Chapters 2 and 6 include details on these integrated alarm and locking systems.

CHAPTER 1

Tactics in Illegal Entry

The outcome of an entry operation depends on two important factors: tactics and techniques. The chosen tactics determine how to conduct the entry as an operation and how to avoid being detected or caught. The techniques are the actual methods used for bypassing locks, doors, and other obstacles, such as alarm systems. This chapter will concentrate on tactics, while the following chapters will detail the actual techniques in use.

The first thing to do when planning an entry operation is to reconnoiter the target (i.e., the actual building or installation to be entered) and the area around it. This is the reconnaissance phase.

Target Reconnaissance and Initial Planning

First of all, the following factors about the target must be determined.

1) Protection:

- number of guards, if any
- location of guards at particular times
- their equipment and armament
- technical details on alarm and controlled-entry systems, if any

2) Layout:

• number and locations of entrances and exits (both ordinary and emergency exits) and hallways or stairways suitable for a quick escape (and possibly windows, rooftops, and sewers—even chutes in an emergency)

• location of important offices or other rooms of importance

• method of smuggling entry team and equipment into the building if ordinary access is denied

• possibilities of temporarily cutting off the entire building from the telephone network and other ways of preventing communication with the outside, for instance, with cellular telephones or radio transmitters

3) Personnel:

• number of staff in the building during and after ordinary office hours

• their location at particular times

4) Timing:

• when the target is most vulnerable

• if there are any outside factors which will influence the accomplishment of the entry operation

Most targets are most vulnerable between 2 A.M. and 4 A.M., as most people are asleep at that time, and any security guards tend to be sleepy and therefore less vigilant during these hours. It is, however, vital to check the guards, as different individuals keep different habits.

There are many types of outside factors which might affect the mission. If breaking into a government building or an industrial complex, especially one of military importance, consider, for instance, the possibility of a surprise inspection by a senior official surrounded by a large number of guards. This is perhaps a worst-case scenario, but other things may also affect the operation, such as the sudden removal or installation of important machines. Such work, and also regular maintenance work, is often performed outside ordinary office hours so as not to delay the regular work.

The vicinity of the target building must also be carefully reconnoitered. Suitable escape routes must be prepared, both for use after a successful entry and in the case of failure. If several operatives are available, it is advisable to organize an outer security ring around the target during the actual entry. The members of the outer security ring will be positioned some distance away from the actual target. They will observe access routes and warn of the approach of police or enemies, either on foot or in vehicles. Warnings are generally passed on by radio, preferably through the use of innocent-sounding code words, so that the transmission cannot be identified if it is monitored.

One way of obtaining information on the tenants of a particular building is to play the role of a private detective. Such detectives sometimes pry into the comings and goings of tenants, especially in divorce cases. This is more common in certain countries than in others, however, so operatives should be careful when assuming this role.

Another way of getting information on tenants is to go through their trash. An office especially yields a surprisingly large amount of trash, most of it in the form of discarded documents. This can give much valuable information. This method is highly useful in determining whether an entry operation is called for or not.

Torn or burned scraps of letters might be found that will indicate whether an entry and a search will be fruitful or not. Burned or dirty pieces of paper can be read through the use of special equipment. It is not unheard of for a businessman to very carefully destroy his important letters, while his secretary simply discards her stenographic notebooks in the wastepaper basket, where they are found by a delighted operative who is trying to piece together the businessman's activities.

At this stage, it is also prudent that every member of the entry team familiarize himself with the appearance of everybody working in the building at the proposed time of entry, as well as that of every resident or employee of the target apartment or office. In this way, the members of the outer security ring can easily recognize any potential threat and sound the alarm in time to warn the entry team if somebody is approaching the target area.

This familiarization process has frequently been accomplished by sending the members of the team to the building in daytime, dressed as cleaners, maintenance crew, or repairmen. They can then spend plenty of time fixing some minor problem or painting a wall or another object, and at the same time get a close look at the entire staff of the building. One good way of identifying everybody is to litter the floor with tools and then warn every passer-by not to stumble on them. Of course, a sufficient cover is required for this kind of job, including the address and telephone number of the company which is supposed to have sent the repairmen there, and somebody on the premises who has called for them in the first place. The name of the company should also be displayed prominently on both their car and the boxes or bags containing tools the repairmen bring into the building. The normal amount of traffic in the building at the proposed hour of entry must also be determined carefully. The behavior and patrol pattern, if any, of the watchman or the security guards must be noted in particular and compensated for in the final entry plan.

The reconnaissance phase is followed by the planning phase. First of all, it must be decided whether the entry is to be covert or overt.

A covert entry is generally preferred, as a successful covert entry will guarantee that the enemy does not realize his secrets have been exposed. The disadvantage, however, is that a covert entry is difficult—sometimes impossible—to execute, and there is a much greater chance of failure.

A covert entry can also be used by police to acquire information without having the proper authority to enter the target building.

An overt entry, on the other hand, is much easier to perform. The disadvantage is, of course, that the enemy will be alerted to the fact that an entry has occurred. Therefore, in order to minimize the chance of his finding out who executed the entry, as well as to maximize his concern that vital secrets have in fact been lost, the entry is frequently camouflaged as a simple burglary, executed for profit.

In either case, the planning phase consists of more or less the same type of work. However, it must be realized that the planning described in this chapter might take several weeks. An entry operation is consequently not to be undertaken on the spur of the moment.

First of all, if the target is located in an apartment complex, it is necessary to decide how to enter the building that contains the target office or apartment. This is frequently done by renting another apartment in the same building. Sometimes it is instead possible to enlist the help of the owner or superintendent, or some other worker or resident in the building. The superintendent is the preferable choice, however, as he frequently has a master key to all apartments in his building. In either case, the personal characteristics and loyalty of the helper must be researched thoroughl, so as to preclude a security leak at a later stage. If the man turns out not to be trustworthy, or none of these alternatives are viable, the team must find a plan to both enter the building and then to perform the actual break-in.

If the target is a villa, small house, or mansion, the procedure is more or less the same. A neighbor might be available, or the entry team must get into the neigborhood and perform the actual entry at the same time.

If it is impossible to secure a master key, it might still be possible to obtain and make a copy of the relevant key prior to the operation. One way of doing this is to search the owner of the key's pockets at a time when he is unable to notice the search. Such opportunities are not easy to come by but might arise in public baths or gyms. A key is not just useful when entering the premises; it is also invaluable if the operatives have to make a quick escape.

As was noted above, a sufficient number of escape routes must be prepared. If a master key or the actual key to the target area can be secured, a sufficient number of copies should be made to ensure that no member of the entry team will be trapped in the building in case of an emergency. This might be a vital precaution.

The members of the security ring must also be experienced enough to delay any employee or resident of the target apartment who might want to enter the premises when the entry team is working. Such a delay can be caused by a member of the security ring pretending to be drunk and accosting the employee until the entry team has managed to withdraw. Another possibility is pretending to be involved in maintenance of the elevator. Minor car accidents or similar situations might also be used to effect a delay. Initiative is very much in demand here, as nobody can plan for every possible situation.

An inner security ring is sometimes also organized specifically to deal with such situations. It is not always necessary to leave the building, even if the security ring gives warning, however. Sometimes the entry team can simply retreat temporarily to a nearby office or apartment and await the moment when they can resume their work. This option is generally not available during overt entry operations, however.

Every member of the entry team should be supplied with a convincing and documented cover story explaining what he is doing on or near the premises. The contents of his pockets and his clothes must conform to the cover. If caught on the premises, he must at least manage to convince his captors that he is an ordinary burglar. The equipment brought onto the premises, as well as all bags, must also conform to the cover if at all possible.

As a final note, it is prudent to rehearse the entry several times in a safe location so that every team member knows exactly what he is to do and how he will do it. The actual entry operation should, if possible, be performed without talking or even discussing what to do, as long as the team is in hostile territory or in the target area. Remember that a chain is only as strong as its weakest link, and this also applies to an entry team. Only if every man knows exactly what he is supposed to do will the operation proceed smoothly.

An overt entry is performed by forced means in which the operatives completely disregard the fact that the entry will definitely be noticed. The team will simply break any locks, doors, or safes standing in the way using the easiest method at hand, without bothering with such niceties as picking locks. Alarm systems will be evaded if possible, or else disregarded. Speed and force are more important than surreptitiousness.

Alarm systems, if they merely alert a distant security company or a police station, can often be disregarded completely when using this method of entry. The reason for this is that the alerted security team or police will not proceed fast enough to reach the penetrated area in time to catch the operatives. Both police and security companies are reluctant to rush to a site when an alarm system has alerted them because the vast majority of such calls are false alarms caused by defective systems.

The overt entry team essentially acts as a team of burglars. No finesse is involved—only brute force. In some countries, ordinary criminals are actually employed by the intelligence service for this purpose, but this is not recommended, as these individuals generally are not reliable and frequently cannot keep silent about missions executed.

The overt entry team should consist of operatives equipped with crowbars, sledge hammers, and a carborundum wheel with circular saw attachment. This is an extremely hard silicon carbide grinding wheel, excellent for breaking through all sorts of hard steel and other strong materials, and thus very useful in cracking safes. Explosives might also be required, although this is less common nowadays. In the past, explosives were commonly used for safe cracking, but a carborundum wheel is much more efficient and easy to control. Batteries for high-voltage power might also be necessary if the fuses in the building blow because the power load is too high or electrical power is unavailable for some other reason.

When planning an overt entry, it should be remembered that the normal means of entry—the front door, for instance—is frequently more difficult to break than some other part of the building, such as a roof or a wall. The former is often reinforced, which very seldom is the case with other parts of the building, at least in smaller houses and villas.

Covert Entry

A covert entry is much more difficult to perform than an overt entry. Speed is important in this operation, too, as every minute in the area of operations brings a chance of discovery and capture. But even more important are skill and diligence. Every lock must be picked, and the means of entry must never leave any marks revealing that an entry has occurred. In fact, every part of the premises must be left in exactly the same condition as it was before the entry. As this requires considerable skill, it should not be attempted by inexperienced operatives; the chance of the enemy noticing the entry is simply too high. Furthermore, a covert entry must frequently be called off because a certain lock or safe proves impossible to pick open in the time available to the operatives.

When all the details of the planning phase are taken care of, a preliminary entry is sometimes undertaken. This entry is executed very cautiously, however, and no attempts are made to find or obtain any interesting documents in the target area. Instead, the sole purpose of the preliminary entry is to make certain that the plan is valid and sufficient expertise is available to perform the actual entry operation. At the time of actual entry, the team and its equipment are generally too conspicuous to allow for failure and the subsequent chance of exposure.

The preliminary entry must be made in absolute silence and with extreme caution. This is because voice-activated tape recorders might be hidden in the target area, and there might be any number of innocent-looking traps, such as short lengths of thread or hair, paper clips, books and papers, or other small objects positioned in a certain spot or arranged in a special way. Such traps are set up to alert the target that somebody has tampered with his things or entered the premises. Before any object is moved, it is therefore necessary to take accurate measurements of the position of every object. This is most easily done with the help of a Polaroid camera.

There might also be more devious traps, such as a video camera and recorder initiated by a sensor capable of detecting an individual on the premises.

The operatives must also remember that many computers will register the time at which they are turned on or execute a command. It is therefore very risky to check the files of a computer, unless the operative knows exactly what he is doing. (Of course, this is never a problem with forced entry operations. As computers are valuable objects, the entire set can be taken easily, without anybody wondering why the burglars stole it.)

Yet another problem is areas covered by dust. The operative must not remove an object from such an area unless he can replace the dust in a convincing way. A small atomizer filled with talcum powder mixed with powdered charcoal can sometimes be used to simulate dust.

During the preliminary entry, the operative should also make a detailed check of the premises. He might find, for instance, that there will be a need for blackout curtains during the actual entry operation. The operative must then determine exactly how many such curtains are called for, as well as their required size and a method of attaching them to the windows. Another point to consider at this time is the number and types of safes and file cabinets. As many details as possible on them, as well as on all ordinary locks in doors and so forth, must be collected, including any numbers on the locks. It is frequently necessary for the locksmiths to know these details in advance in order to pick the locks successfully during the actual entry operation.

It is also important to find a safe place to put camera equipment and a small, portable Xerox machine during the actual operation. Preferably, this will not be inside the office or apartment to be searched, as it could preclude or hinder a quick escape. A small, nearby cleaning storage room is ideal for this purpose. Rest rooms should be avoided, too, as they might be frequented by a person making an unexpected call to his office or apartment during the entry operation. If the chosen camera site is sufficiently remote, it might even be possible to return later to recover the equipment hidden there if a quick escape is required.

The successful execution of a covert entry requires very careful planning as well as considerable skill. In addition, it is necessary to maintain total security from interruption during the search of the target building. This necessitates the use of both an outer security ring and probably an inner security team able to delay any intruders.

A covert entry team must consist of several individuals, each an expert in his field. Among them should be a lock-picking expert, a safe expert, and one or more experts on alarm systems in order to execute the actual entry. The alarm system experts must be prepared to manipulate electronic locks and other electronic systems such as lift machinery.

Furthermore, one or more analysts capable of rapidly evaluating any found documents and at least one photographer should be present. The photographer must be equipped with several cameras, both for copying any found documents, and, equally important, snapping Polaroid photographs of the original appearance of the rooms to be searched, in order to provide a pattern for restoring everything to its original place before leaving the premises.

Nowadays, as was noted above, it is common to use a small, portable Xerox machine to duplicate any found documents. Of course, a sufficient supply of film, paper, and so on must be included in the equipment brought onto the premises. Sometimes infrared photography techniques can be used, as these require no visible light source.

There should also be an expert in opening letters, capable of opening and resealing any type of letter in a convincing way. This technique also requires a certain amount of equipment. It might even be required to bring an infrared fluorescence detection system in order to detect traps on sealed letters or documents, or alterations to passports or other documents.

Finally, any guns or other means of dealing with individuals interrupting or noticing the entry must be decided upon. Every piece of equipment, including radio transmitters, must be tested and checked so that nothing is found missing after the team has entered the premises. Radio transmitters in particular might require a test in the target area before the actual operation is initiated. The range of radio transmitters depends very much upon the construction of the building, as different buildings disturb radio transmissions in different ways. It might also be a good idea to bring a spare camera; a broken camera in the middle of an operation is a bad excuse to pack up and go home.

During the actual entry, the premises should be entered by only one individual operative. This is to ensure that the entire group does not walk into a trap. When this operative gives the agreed upon signal, the rest of the group will follow him in.

There are several ways to prepare for a covert entry operation. One good way, already discussed above, is to secure a master key. This is sometimes possible in apartments and office complexes, as the caretaker generally needs one. The same is true of hotels. However, it is generally impossible in villas and mansions owned by private individuals.

Another obstacle might be an alarm system. Such systems, which frequently sound the alarm because of technical defects, will not be taken seriously by investigating police patrols or private security companies, however. It is therefore possible, at frequent time intervals over a period of several days or weeks, to alert the alarm system discreetly and then quickly withdraw from the area without leaving any traces. Then, during the actual covert entry, the police will be much less vigilant. This allows the entry team a longer period of time to do what they came for.

This might also be true of an ordinary guard. During the early years of the Second World War, Willis George worked for the Office of Naval Intelligence. As was mentioned above, he planned and executed several entry operations during these years.

During one of these, he found out that a foreign consul had posted an armed guard in the private part of the consular office. Obviously, this made a covert search impossible. George therefore had to devise a means of getting rid of the guard without exposing his own operation. The guard had probably been posted there because the consulate elevator operator had become suspicious during a previous covert search in the building.

George finally decided to try to make the elevator operator appear ridiculous by causing him to behave in a seemingly overzealous manner. If the consul eventually decided that the guard was unnecessary, then he would likely dispense with such extreme measures for protecting his office.

With this in mind, George once again entered the building for the sole purpose of deliberately making some noise in order to alert the elevator operator. Then he hurriedly left the scene.

About half an hour later, the consul arrived by taxi. The elevator operator had, just as was expected, called for him. They searched the office, of course without finding any signs of an entry. The consul was angry and left after the fruitless search.

A few nights later, George repeated what he had done. Once again, upon hearing the suspicious noise the unfortunate elevator operator called the consul immediately. The consul arrived, angry and tired of being awakened every other night. Once again, their search produced no indications at all that an entry had been attempted.

The next night, as the guard was no longer posted in the office, George and his team were able to continue preparing for the actual entry.

In this context, it should also be noted that the majority of all false alarms take place in the morning between seven and nine and in the evening between five and seven. The reason for this is that these are the times when the ordinary occupants of the buildings turn the alarm systems off or on.

The police and serious security companies know about this and are therefore less vigilant during these hours.

When on the premises during an illegal entry, the operatives must think of many things. They must not, for instance, use any toilets or water faucets. This could leave traces and make noise in the plumbing pipes. Smoking is also prohibited, of course, as this leaves both traces and odor.

Finally, when leaving the premises after a concluded operation, whether successful or not, it is important to check that no piece of equipment has been left behind and that the appearance of the target building is exactly the same as before the entry. This might require polishing or even rewaxing the floor, if the operatives have entered with their shoes on. All fingerprints must be wiped clean, whether on doors, walls, or office equipment. If a safe has been opened, the dial must be reset at its original reading. If the apartment contains thick rugs on the floor, it might even be necessary to sweep them upon leaving so as not to leave any footprints.

It is necessary to realize that every situation is different from any other, so while the rules above are useful, there might be other things to think of, too. Only extreme diligence and caution can make a covert entry remain so.

CHAPTER 2

Locks and Lock Picking

Lock picking can be defined as the method of opening a lock mechanism by the intrusion of special tools other than the regular key. Lock picking is not easy. It is, however, possible to open any lock without the proper key, but some locks are definitely more difficult to open than others. In this chapter, we will look at the basic types of locking devices and how they can be opened most efficiently without having access to the proper keys.

The reason lock picking is at all possible is that there are always certain tolerances built into the design of the lock. The various parts of the mechanism never fit perfectly. There will always be some diminutive empty space in which the lock-picking tools can be inserted.

An expensive lock is usually, but by no means always, designed and manufactured with less tolerances than a cheaper one. Less tolerances mean less space to insert lock-picking tools. A cheap lock is therefore almost always easier to open, as there will be tolerances large enough to insert whatever tools are required to pick the lock. No lock is absolutely pick-proof, although today there are numerous types that are extremely difficult to open. In some cases, the level of difficulty is so high that the lock is effectively impossible to pick under field conditions.

However, as the price of the lock dictates the quantity

that can be sold, the vast majority of locks are fairly cheap and consequently easier to pick open. For this reason, there is a very good chance that the lock encountered during a mission will be one of the cheaper types.

The devices used as locks today can be divided into the following general types:

- warded locks
- lever tumbler locks
- disc tumbler locks
- pin tumbler locks
- tubular cylinder locks
- magnetic locks
- combination locks

Locks of the types mentioned above appear in the following shapes:

- luggage locks
- padlocks
- vehicle locks
- mortise locks
- surface-mounted auxiliary locks

The last two categories are the two general shapes of door locks. They can be found all over the world. Although these general types contain numerous design variations, all locks encountered will fall within one of these groups, with the exception of those that represent a melding of two different types of locking devices.

Luggage locks, pad-

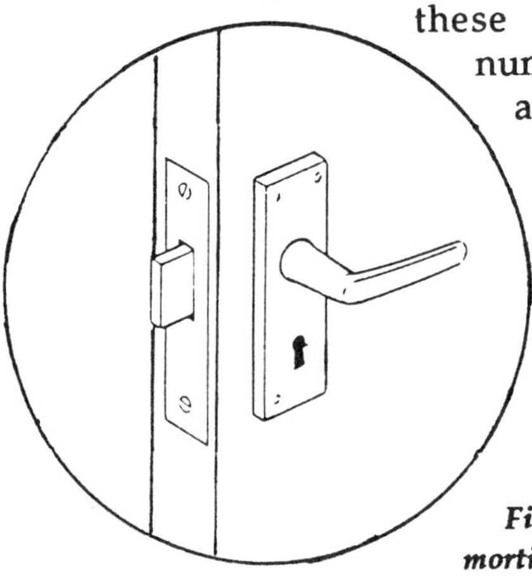

Figure 1. Rim lock (left) and mortise lock (right).

locks, and vehicle locks will be described fully later in this chapter. The door locks need to be described in more detail, however, as you are most likely to encounter these in the field.

There are basically two types of external door locks—rim locks and mortise locks (fig. 1). Rim locks are surface-mounted, screwed to the surface of a door. They usually have a spring-operated, beveled latch bolt that automatically springs back when the door is shut to hold it closed. The door is opened by turning back the latch using a key or an internal knob, or by moving a sliding handle. In most rim locks, the latch bolt is checked by a safety catch so that the door can also be shut without latching. Rim locks are also known as locking bodies.

Mortise locks mount inside the door, fitting into a mortise, or slot, in the door's leading edge. They are therefore very neat and slightly stronger. However, it is generally the wood rather than the lock that gives way during a break-in, so this means little. A rim lock is therefore more commonly fitted to a thin door to prevent the door from being weakened. For the same reason, the staple of a rim lock will not weaken the door frame, as it is surface-mounted. However, the staple itself is only held by screws and can be broken away from the frame.

The simplest rim lock is the night latch. This is an auxiliary lock with a spring latch bolt that holds the door closed. The night latch functions independently of the regular lock on the door and cannot be deadlocked. The latch is operated by a key from the outside and the knob from the inside. In this lock, the spring bolt can be pushed back with a piece of flexible plastic. (For instance, see the section on transom entry, doors, and door chains in

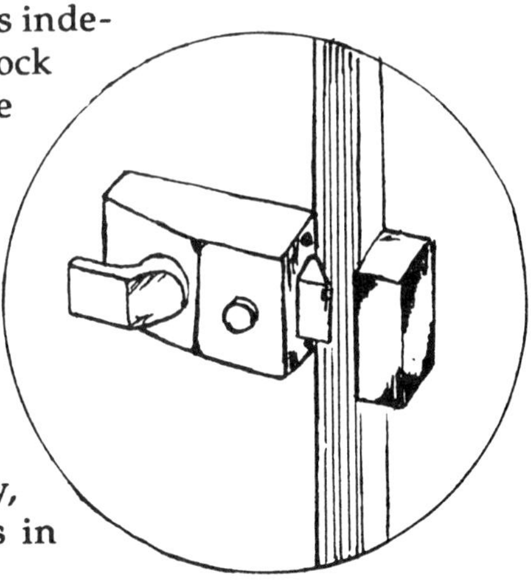

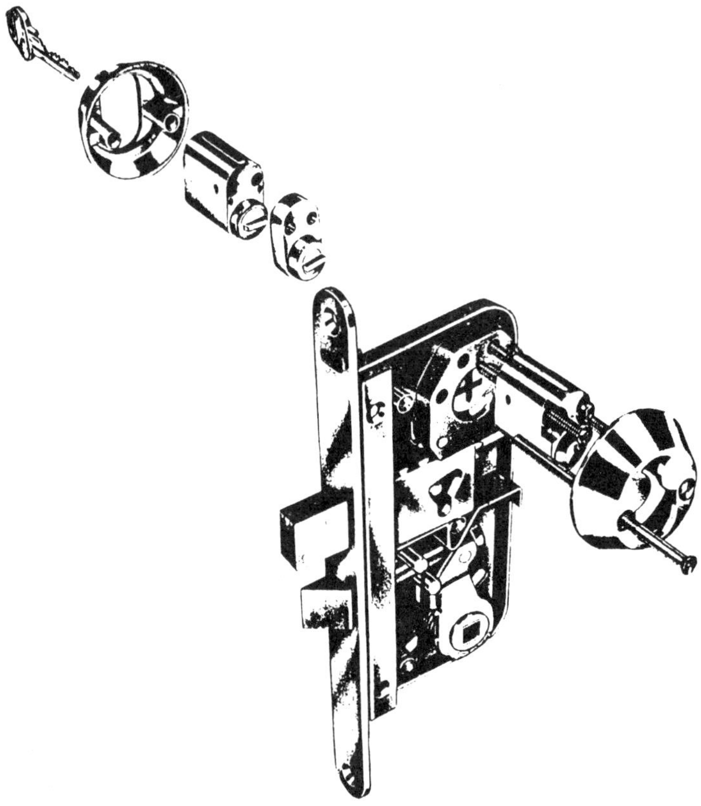

Chapter 4.) Another method is to break a small adjacent window in order to reach through and open the lock from the inside.

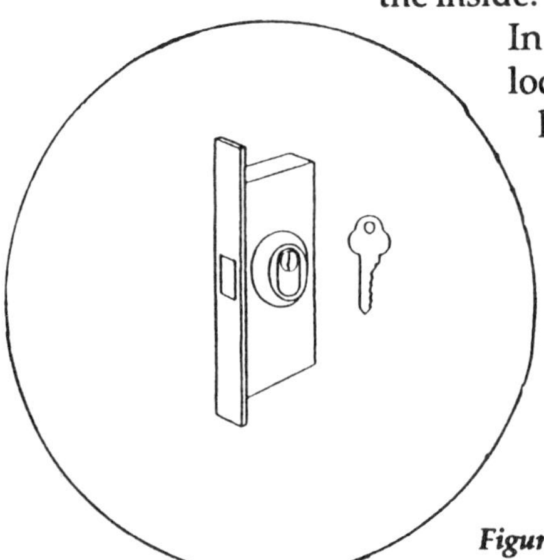

In practice, most simple rim locks nowadays have a deadlocking mechanism, operated by a small thumb piece from the inside, which prevents the spring bolt from being forced back.

There are also other types of surface-mounted

Figure 2. Cylinder lock.

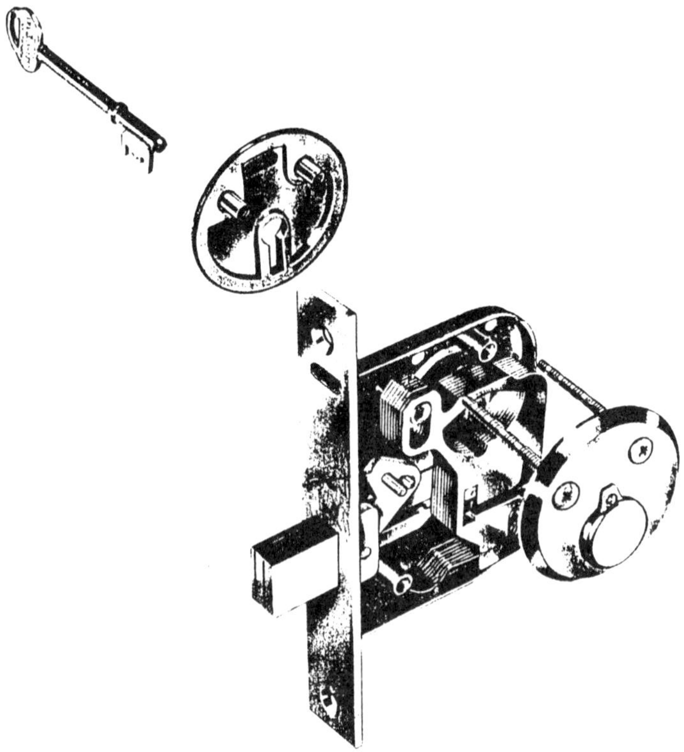

auxiliary locks. They include dead latches (locks that can be automatically or manually locked against end pressure when projected) and surface-mounted cylinder locks used separately from another lock unit. Door chains, surface bolts, and chain bolts are also usually counted as surface-mounted auxiliary locks and will be described in later chapters.

More advanced rim locks, such as the ones with an automatically deadlocking latch (automatic deadlock) or a manually deadlockable latch, always have a spring

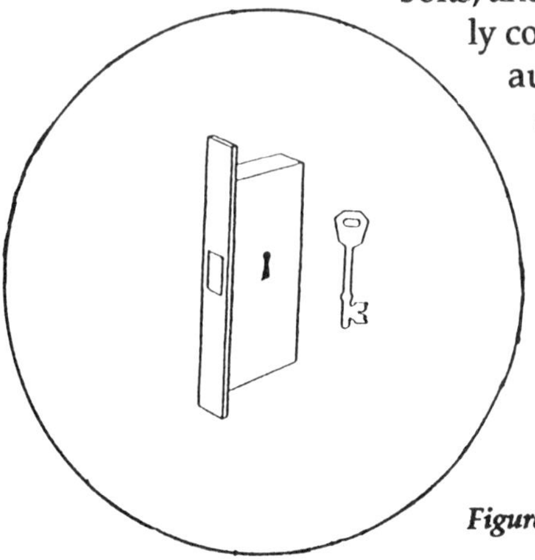

Figure 3. Lever lock.

bolt with a mechanism that prevents the bolt from being forced back when the door is shut. The internal handle can then be deadlocked by using a key, either from the inside or the outside. The lock cannot then be opened except with a key or by picking it.

Rim locks are usually operated by pin tumbler cylinders. A mortise lock, however, is either operated by a pin tumbler cylinder (fig. 2) or by flat levers in the lock case itself (fig. 3). The levers are of different heights to correspond to the cuts in the key. The number of lever tumblers determines the security level of the lock. The more levers there are, the greater the number of key variations and the more secure the lock. Five-lever locks are common, but in many countries the nine-lever lock is considered the standard today.

Some mortise locks also rely on cylinders, if the owner wants to have the same key for operating both the front and the back door, for instance. Lever-type mortise locks can also be designed to have identical keys, of course.

No part of the body of a mortise lock is visible when the door is closed, as it is concealed within the door. For this reason, the mortise lock is generally not fitted to doors less than 44 millimeters thick.

There are basically two types of mortise locks: the key-operated dead bolt and the two-bolt mortise lock, or sash bolt.

A dead bolt is a lock bolt that has no spring action. For this reason, it is always actuated by a key or a turn knob. This lock is a true deadlock. A sash bolt has both a latch bolt and a dead bolt. The latch bolt is a beveled spring bolt that is operated from either side by the door handle, while the dead bolt is operated by the key. Two-bolt mortise locks are often fitted on back and side doors, while key-operated dead bolts are fitted to front doors. The dead bolt can be operated from the inside by a thumb piece.

A deadlocked mortise lock cannot be opened from the inside without a key if it has been locked from the outside. The reason is that the bolt cannot be withdrawn into the lock case unless the key is used or the lock is picked.

Warded locks were first invented by the ancient Romans. The warded lock relies on one or more wards to protect the internal lock mechanism. A ward is a protruding ridge in a lock or on a key designed to permit only the correct key to be inserted in the lock. Warded locks are of a fairly simple design and can be found all over the world. They are still used in door locks in older areas of even metropolitan cities such as New York, despite the fact that they are very insecure. They are also common in old padlocks. Student locksmiths frequently use these locks for practicing lock picking.

Warded door locks are either of the rim or mortise type. Both types of locks operate on the same principle. The surface-mounted rim lock is generally even less secure than the mortise lock.

Normally there are two interior wards in the lock, positioned directly across from each other. One is on the inside of the cover, while the other is on the inside of the backing plate.

The key for a warded lock is cut to correspond to the single or multiple wards that have been designed in the lock. The key will only come in contact with the actual locking mechanism after it has passed all of the wards. Then the cuts on the key will lift the lever to the correct height and throw the dead bolt into the locked or unlocked position. As long as the dead bolt is retracted, turning the doorknob will activate the spindle and release the door.

The wards are of three possible types. One type of side ward is designed to allow only a key with a slot milled on the edge to pass (fig. 4). Another type is designed instead to allow only a key with the slot milled on the

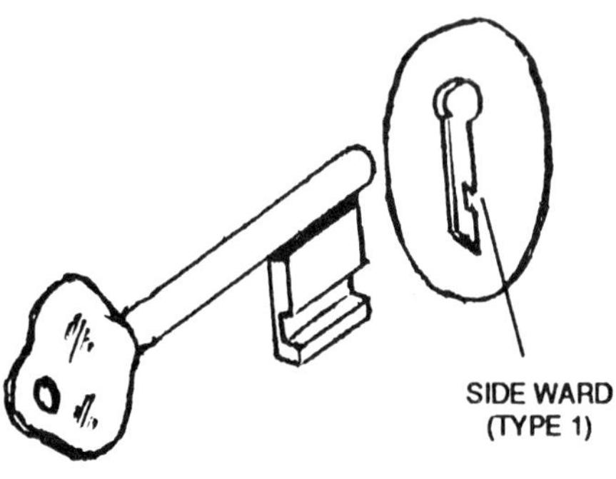

Figure 4. One type of side ward.

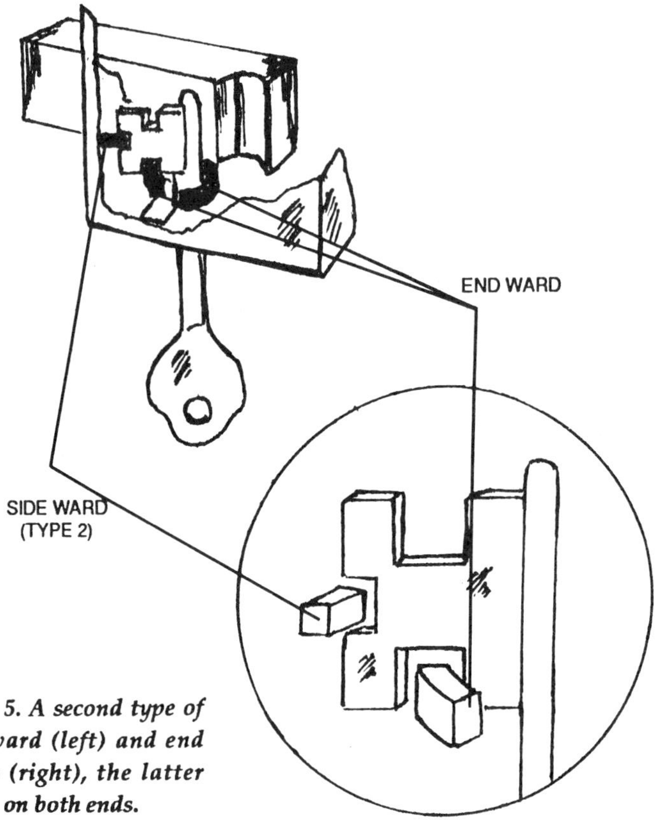

Figure 5. A second type of side ward (left) and end wards (right), the latter milled on both ends.

side of the key to pass (fig. 5). An end ward, finally, will only allow a key with a slot milled on the end to pass (fig. 5). End wards are commonly milled on both ends, as the key then can be used from both sides of the lock.

Side wards, at least, can generally be passed successfully by inserting a skeleton key—a key that has been ground down on the sides to become thin enough to bypass these wards (fig. 6). Such a skeleton key can be helpful in opening the lock, although this is by no means assured. Skeleton keys can be bought in variety stores, but generally not from reputable locksmiths. A properly made skeleton key will operate almost any warded lock whose keyway accepts it.

All the ward cuts in a skeleton key have been opened up so that only the tip that is necessary to operate the latch spring remains. Most warded locks can also be

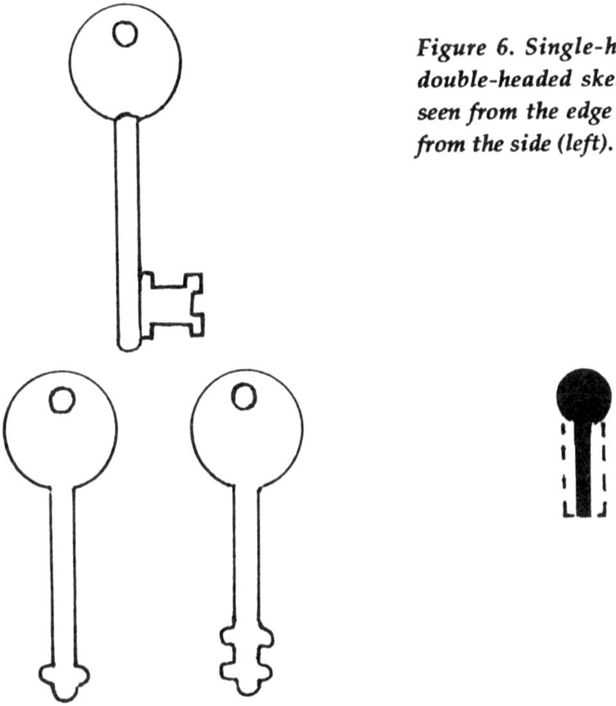

Figure 6. Single-headed and double-headed skeleton keys, seen from the edge (right) and from the side (left).

picked with a T-shaped lock pick, of course.

Nowadays many lock manufacturers try to raise the security level of their warded locks by adding another spring latch with a ward between them. Then a double-headed skeleton key can be used. The principle remains the same, but this skeleton key is designed to handle that extra complication as well.

A good locksmith with plenty of time can make a duplicate key to the lock using the technique known as impressioning. This is the method of determining the shape of the key by simply studying the lock from the outside. The locksmith will insert a key blank smoked by a candle into the lock. A key blank is a key that has not yet been cut or shaped to operate a specific lock. The smoked key blank, when extruded, will show several small marks where the candle black has been removed. These marks will tell the locksmith what cuts to make, where to make them, and how deep they must be. This is a fairly lengthy process, and it requires some skill. Nevertheless, it is sometimes extremely useful, as the process will provide the entry team with a true copy of the key they need to use.

Because of their simplicity, warded locks are very easy to pick. Sometimes a pair of wires will be sufficient to use as lock picks. In that case, one of the wires will be used for throwing the bolt, while the other is used for adjusting the lock mechanism to the proper height for the bolt to be moved, if this is required.

The main difficulty in picking a warded lock is not to negotiate the few wards that are obstructing the pick, but to find the correct set of lock picks. Here it is important to have picks of the correct size. As was previously mentioned, skeleton keys are often easier to use. Precut blank keys are therefore often used for this purpose instead of regular lock picks.

Lever Tumbler Locks

The lever tumbler lock, or lever lock, was first introduced in the eighteenth century. These locks are still common in light security roles today. They are often found on desks, lockers, mailboxes, bank deposit boxes, and similar objects. However, now a modified, much more pick-resistant variety of the lever tumbler lock is also in worldwide use as a high-security mortise lock. In this case, the lock might use as many as nine levers or more.

It is very important to realize that although the security level of the minor lever tumbler locks is lower than, for instance, the pin tumbler locks described below, the security level of the mortised lever tumbler locks is generally significantly higher. These locks are difficult to pick.

A lever lock (fig. 7) consists of six basic parts. These are the cover boss, the cover, the trunnion, the lever tumblers (usually two, three, five, but sometimes six, twelve, or even fourteen in deposit box locks), the bolt, and the base. The lock is operated by a standard flat key. After the key has been inserted into the lock, the key is turned, which causes the key cuts to raise the level tumblers to the correct height.

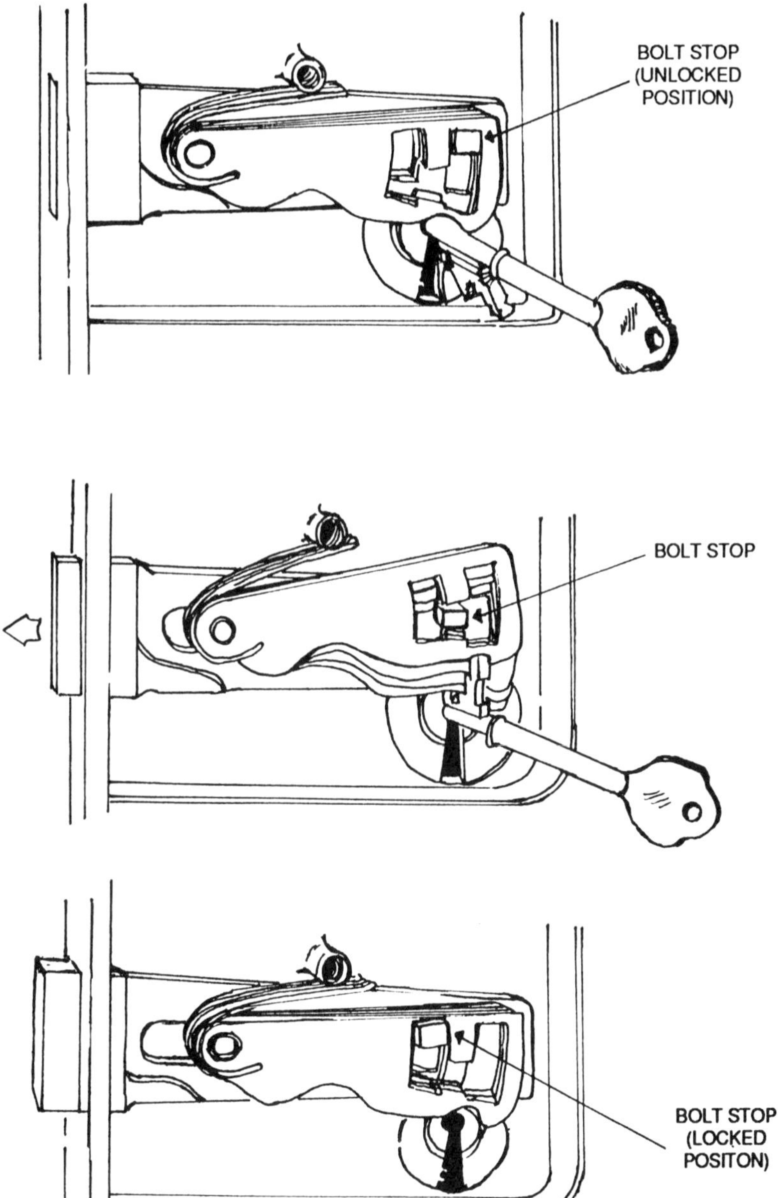

Figure 7. Lever tumbler lock.

As the levers are raised to the correct position, the gates of the lever tumblers will align and release the bolt. The bolt stop is allowed to pass through the gates from the rear to the front or vice versa; this will unlock or lock the lock.

The lever tumbler gates must be perfectly aligned, or the lock will not function. This enhances security, as the key must be cut perfectly.

The key to a lever lock is almost invariably flat. Here, too, an experienced locksmith can make a new key by impressioning. The process is much more difficult, however, than was the case with the warded lock. First of all, the locksmith must do what is generally known as "reading the lock." With a reading tool—simply a slightly bent (for a view of the tumblers), stiff length of wire, about 7 or 8 centimeters long, with a wooden handle to make it easier to hold—he can probe the narrow lock keyway. This will give him some idea of how to cut and shape the key.

The locksmith will then use the positions of the lever saddles, the part of the lever tumbler that is in direct contact with the key, as one clue to the design of the lock. The wider the saddle, the deeper the cut on the key. This process takes considerable skill and long practice. There are also locks that apparently have the same saddle width on each lever. These are even more difficult to read. Here the locksmith must attempt to determine the design by finding out how high he can raise the various levers.

In order to pick a lever tumbler lock, begin by inserting the torque (or tension) wrench. This is a special device used to apply pressure on a lock while its tumblers are being manipulated with the pick.

Push the wrench to the lowest point within the keyway, as this will give the pick maximum work space. Locate the key notch in the underside of the bolt and apply pressure (fig. 8). The bolt stop, which is affixed to the bolt, will now bring pressure on the tumblers. By exerting pressure on the lever tumblers with the torque wrench, you can manipulate them with the pick. Therefore, insert the pick into the keyway. The levers must be moved into position for the bolt stop to move through the lever tumblers' gates.

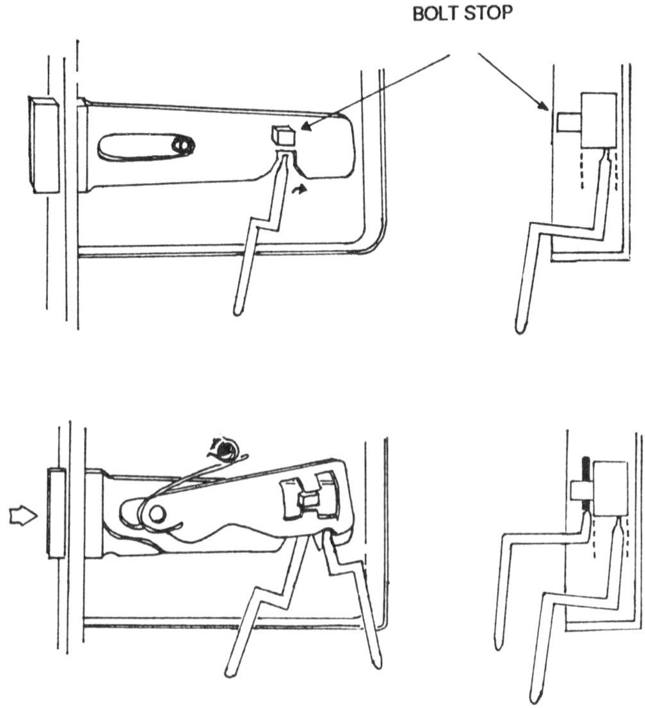

Figure 8. Picking the lever tumbler lock.

One tumbler tends to take up most of the tension, and this is the one to work on first. When this tumbler is raised to the right position for the bolt stop to pass through the gate (not too high, or it will be impossible), you will feel the tension from the bolt slacken through your wrench as the bolt attempts to force its way into the gates. This is the point at which to stop, and then repeat the process with the lever tumbler next to the raised one. When all levers have reached this point, you can get the bolt to pass through the gates by shifting the torque wrench against it. This will open the lock.

If the pressure from the torque wrench is relaxed at any time during the process, all levers in the raised position will drop back to their original positions. Therefore, always keep some pressure on. However, take care not to raise the lever tumblers too high, or they will be raised above the unlocked position. Allow the tumbler to retain its drag as it is raised, as this will help you feel through your torque wrench when you have reached the right position.

Simpler, desk-type lever locks have two parts that must be moved in order to open them. Naturally, the lever must be raised, but the bolt must also be operated to open the lock. This can be accomplished most easily with an L-shaped lever pick. Push back the levers and catch the bolt by turning the pick until you find it. It is sometimes helpful to peer into the lock with the help of a flashlight.

Disc Tumbler Locks

Lever and disc tumbler locks are related in design, although they were invented at different times. Disc tumbler locks are commonly used in garage and trailer doors, but also in many types of cabinets, desks, padlocks, older vending machines, and cars. These locks can be recognized by the fact that the first flat disc tumbler can be seen through the keyway. Disc tumbler locks are also sometimes known as wafer tumbler locks because the tumblers are shaped like wafers, or discs.

The common disc tumbler lock is generally as secure as the lever lock, but less secure than a pin tumbler lock. They are similar in appearance and in the broad principle of operation to the pin tumbler locks; however, the internal design is quite different.

The disc tumblers are flat circular or oval-shaped steel stampings that are arranged side by side in slots in a cylinder core, or plug, within the lock (fig. 9). Every disc will

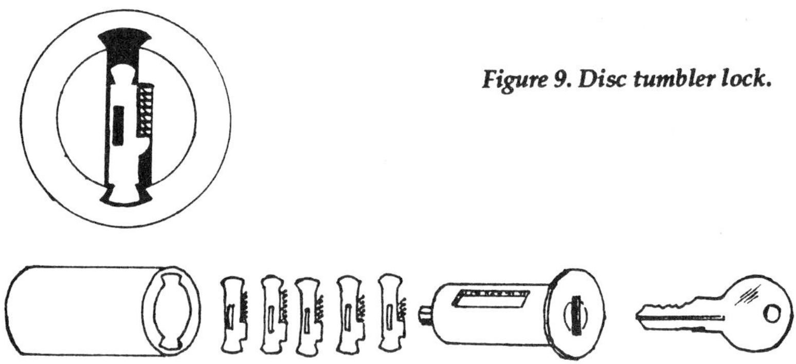

Figure 9. Disc tumbler lock.

have a rectangular cutout in the center, which matches a notch on the key bit. The disc will also have one or more side projections.

This type of lock employs a rotating core, and this is what makes the disc tumbler lock look similar to the pin tumbler lock. The core is cast so that the tumblers protrude through the core and into slots on the inner diameter of the cylinder. As long as the tumblers are in place, the core will be locked to the cylinder. The key, when inserted into the lock, will raise the tumblers high enough to clear the lower cylinder slot. They must not be raised so high as to enter the upper cylinder slot, however, as this will once again lock the plug in position. When the tumblers have been raised to the right position, the plug is free to rotate. This will operate the bolt.

The key to a disc tumbler lock looks like a cylinder pin tumbler key, but it is usually smaller. Furthermore, it will always have five cuts, while a cylinder pin tumbler key might have six or seven. Therefore, the disc tumbler lock is not very secure. As every lock has no more than five tumblers and each tumbler cutout has five possible positions, the design technically allows 3,125 different key changes. In practice, however, some variations are inappropriate, so this leaves us with only around 500 different key changes. Some disc tumbler locks used in offices, for instance in desks, are even simpler, with only 200 possible key variations.

Here, too, an experienced locksmith can make a key by impressioning. First, however, he must read the lock. In the case of disc tumbler locks, this is fairly easy. First of all, you must make a reading tool from a stiff wire. Insert the reading tool into the lock so that you can observe the discs within the lock. Raise and lower each disc by moving the tool until you can see the general positions of all of them. This is generally not difficult, as there are only five variations of the disc tumblers (fig. 10), and their position will give you a general idea of the profile to be used for the key. When this is determined, the key can be impressioned in the usual way by inserting a blackened key.

Disc tumbler locks are often used in offices. Their con-

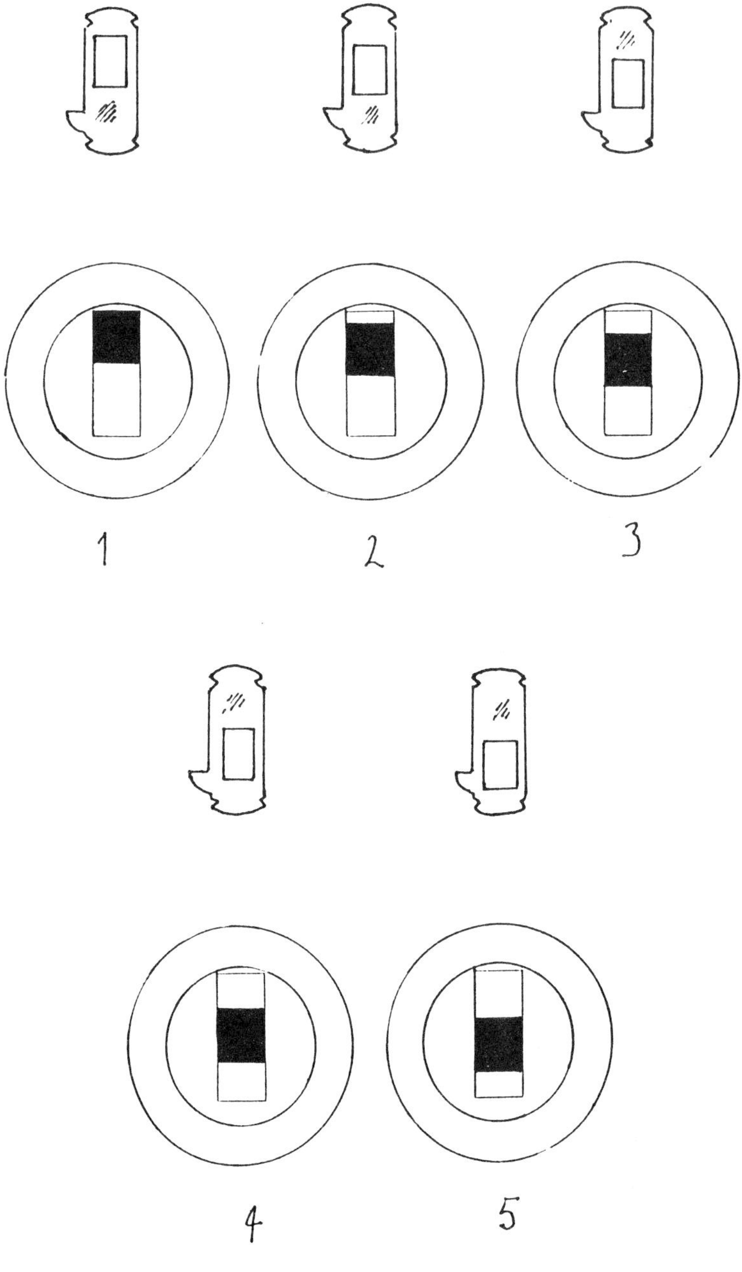

Figure 10. The five disc tumbler variations and their positions in the keyway.

struction is usually fairly simple. Sometimes a simple sliding-bolt lock will be used with the disc tumbler cylinder. In these locks, the bolt is grooved to accept a projection on the back of the plug. The projection engages the groove and converts the rotary motion of the plug into reciprocating motion, opening the lock. In some of these locks (the stronger ones), the bolt-actuating pin is cast as part of the plug (fig. 11). In this case, the plug can usually be released with a probe wire. Locks of this type are often found in drawers and cabinets.

A standard disc tumbler lock (i.e., one that uses a single-sided key) can be opened with the same picks that are required for a pin tumbler lock (these methods will be discussed below). The disc tumbler lock, too, can be picked by bouncing the tumbler to the shear line. The shear line is the space between the cylinder and the plug of the lock cylinder. Usually a rake pick is used for this purpose.

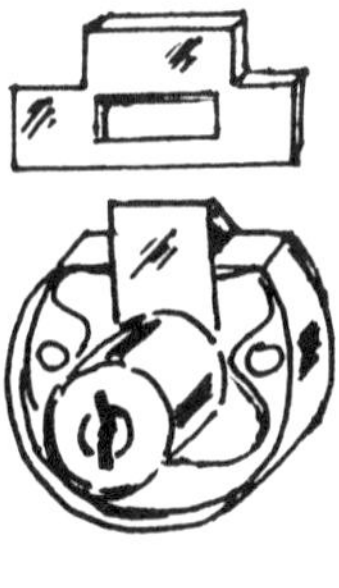

Figure 11. Simple disc tumbler lock.

The bounce method is definitely the best for picking double-sided disc tumbler locks. This is a lock in which there are

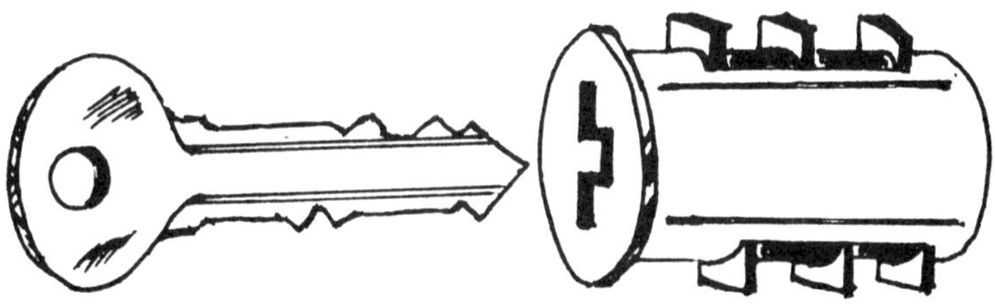

Figure 12. Double-sided disc tumbler lock.

disc tumblers protruding through the core in both sides of the cylinder (fig. 12). Such locks are also fairly common and sometimes require special lock picks. The important thing, however, is not the pick. The picking procedure is the regular one, although it has to be repeated on both sides. After the top disc tumblers have been located and moved to the unlocked position, repeat the process with the bottom ones. Do not forget to insert the torque wrench into the keyway and apply a slight pressure to the core as the pick is pulled out.

Pin Tumbler Cylinder Locks

The pin tumbler lock was first invented in ancient Egypt. The same principle was used much later in the well-known Yale lock, introduced by Linius Yale more than a century ago. Today the pin tumbler lock is one of the most common types of locks in the world. It is used for both residential and office building locks, as well as in numerous other applications.

A pin tumbler cylinder lock is so named because it relies on pin tumblers. Pin tumblers are small sliding pins in the cylinder that work against coil springs and prevent the cylinder plug from rotating until the correct key is inserted in the keyway. Fully assembled, only the plug of the lock (the face of its rotating cylinder) can be seen. Locks of this type are generally more secure than the previously described locking devices. They can be recognized by the first pin that can be seen through the keyway. Even the shear point of the pin can sometimes be seen when looking into the keyway.

The pin tumbler cylinder is a completely self-contained mechanism that can be used with a very large number of lock sets. The basic parts of the pin tumbler cylinder are the cylinder case or shell, the plug or core (the cylindrical mechanism housing the keyway), the keyway, the upper pin chambers, the lower pin chambers, the springs, the drivers or top pins, and the bottom pins (fig. 13). All parts of the cylinder are housed by the cylinder case.

The plug is the part that rotates when the proper key is inserted into the keyway. The drilled holes across the length of the plug can vary in number, but there are usually five or six. Some plugs have as few as four or as many as seven holes. These holes are called the lower pin chambers, as they each hold a bottom pin. The upper pin cham-

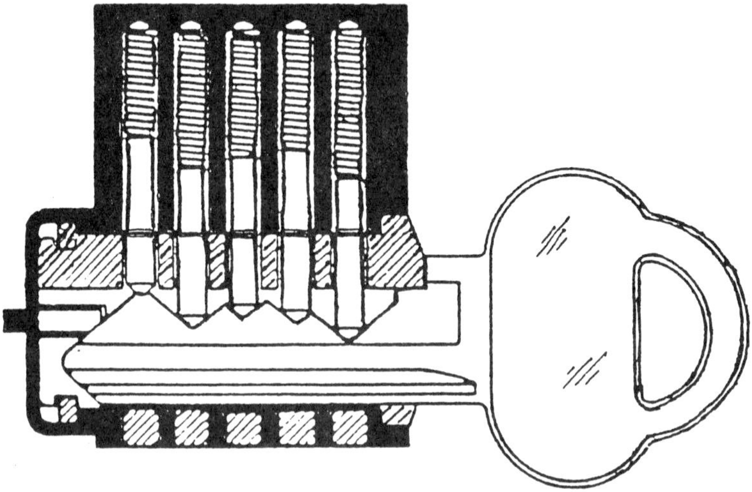

Figure 13. Pin tumbler cylinder lock.

bers are the corresponding drilled holes in the cylinder case directly above the holes in the plug. They each hold a spring and a driver.

The springs and drivers are usually of the same length. The bottom pins, however, are of different length, as they are designed to match the depth of the cuts in the key to the lock by being raised to the shear line by means of the cuts in the key.

Whenever there is no key in the keyway, the springs will press the drivers partially down into the plug so that it will not rotate. As the plug already holds the bottom pins, there is not enough room to allow more than the lower portions of the drivers into the plug.

In order for the plug to be able to rotate, there is a small amount of space between the plug and the cylinder case. This space is called the shear line. When a proper key is inserted, it will force the top of all the bottom pins and the bottom of all the drivers to meet at the shear line. Then, finally, the plug is free to rotate to the open position.

The plug is generally machined with a shoulder at its forward surface, which mates with a recess in the cylinder. If this is not the case, however, it will be possible to open

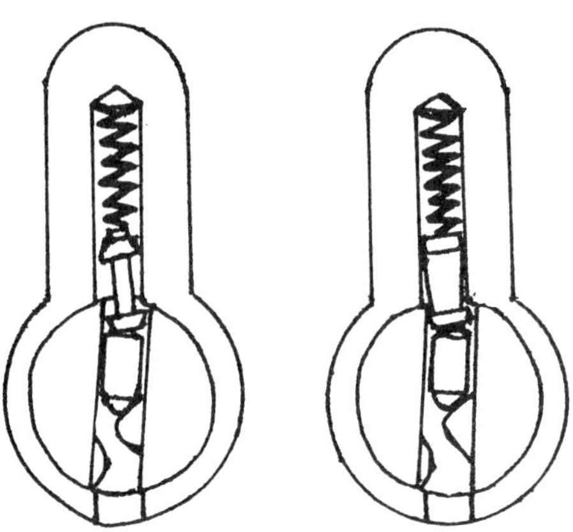

Figure 14. High-security drivers. Spool driver (left) and mushroom driver (right).

the lock by shimming the pins with a strip of spring steel. This would force the pins out of engagement and allow the door to be opened. Contemporary locks do not generally allow this, however.

The pins and the drivers will usually have a broken profile so as to make the lock more difficult to pick. A driver with a broken profile will generally hang up before it passes the shear line. A lock with standard cylindrical drivers is consequently easier to pick. For this reason, so-called mushroom and spool drivers are fairly common in pin tumbler locks (fig. 14). Mushroom drivers are mushroom-shaped. A mushroom driver will interfere with picking the lock, as it will engage with the notched cylinder shell when you attempt to raise the pin to the shear line. The spool driver works in a similar way.

It is also possible to impression a key to a pin tumbler cylinder. The methods are different, however, than for the previously mentioned warded, lever tumbler, and disc tumbler locks. The main difference is that the key blank cannot be smoked, as the soot would wipe off when the key was inserted into a pin tumbler cylinder. The locksmith must therefore depend instead on the small marks left on the key blank itself after it has been exposed to the pin tumblers. For this reason, it is best to polish the key blank thoroughly before it is inserted the first time. The tiny scratches you are looking for will be impossible to see otherwise.

Picking a pin tumbler lock requires the ability to determine when the cylinder pins have reached the shear line. This can be felt through your tool or heard as a minute click.

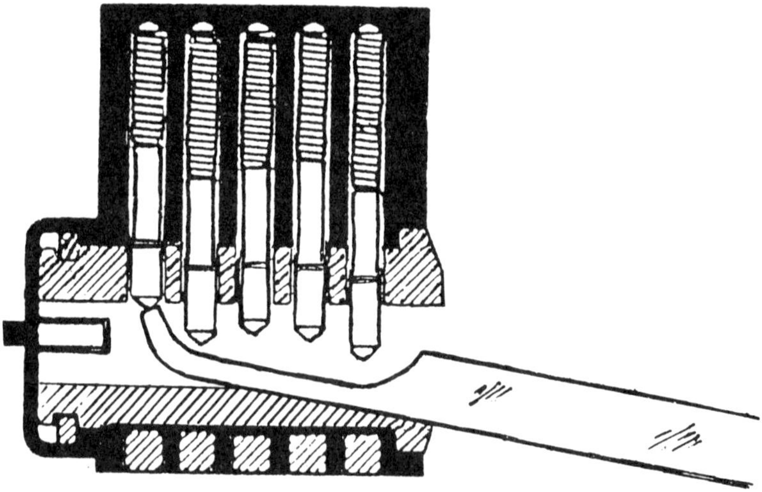

Figure 15. Raising a pin to the shear line.

A feeler pick should be used to raise the pin to the shear line (fig. 15). Be careful not to apply too much pressure, however, as the pick then will raise the pin above the shear line rather than exactly to it. If this happens, the pin will completely block the attempt to pick the lock open. Furthermore, the pin can then easily get stuck at the wrong place.

Before you attempt to raise the pins, insert a torque wrench into the keyway. The reason for this is that moving the wrench slightly to the left or right will hold the drivers tight against the plug. Then insert the pick and, when all the pins are raised to the proper position, use the wrench with just the right amount of pressure to rotate the plug to the unlocked position and open the lock. Never use too much force. Usually only a delicate but firm touch is required to rotate the plug.

It is, of course, difficult to first raise, and then keep all of the pins at the shear line. But the fact that the cylinder pin holes in most locks are not perfectly aligned helps the operative to hold one pin at the shear line with the tension from the wrench while working on the next one with the pick.

The first pin to be raised should preferably be the longest. This is usually the tumbler that takes up most of the tension. Beginning with the longest pin also allows the locksmith to progress from the smallest amount of pick movement up to the greatest. When picking the lock, you will notice that the plug will move slightly for every pin that reaches the shear line. Remember that this movement can also be felt through the torque wrench. This makes it easier to notice when a pin has been raised successfully.

If one of the pins is raised above the shear line, you must release the tension and start again. Less tension is then required when you make the renewed attempt. It is very difficult to judge the amount of pressure necessary to raise the pins in a lock, as even two locks of the same type can react in a totally different way. An experienced locksmith will vary the amount of pressure from light to heavy, depending on what is required. This is more easily said than done, of course, but experience will help.

Mushroom drivers, when encountered, present a special problem. It is easy to describe how to pick these locks in theory, but a large amount of practice is required to do it successfully on a consistent basis. The secret lies in feeling the exact moment at which the driver is engaging the notched cylinder, but before it becomes completely stuck. At this point, slightly release the pressure on the pin before you once again attempt to raise it. With any luck, the driver will have slipped back so that you can now raise it straight up until it is above the mushroom-shaped trap. When the pin is safely raised to the unlocked position, immediately increase the tension on the wrench so that it will not slip down again. Some locksmiths use a spring-loaded wrench for this purpose.

Another way of picking a pin tumbler lock is to use a rake pick or a diamond pick (see Chapter 3) to bounce the pins to the shear line. This process consists of inserting the pick fully and then quickly withdrawing it while applying light tension on the plug. This motion often throws the pins apart because of inertia. The area at the shear line will open up, permitting the plug to rotate. This technique does not work on all types of locks, however.

A rake pick is sometimes also used to rake the lock

open. This is not recommended, as some locks will be damaged. Another point to consider is that merely forcing the pick rapidly in and out of the cylinder (raking) will only bounce the pins above the shear line. Delicacy is required, so if you do not have sufficient skill to pick the lock open, I would strongly recommend bouncing instead of raking.

A very worn cylinder, especially one with loose plugs, is frequently quite easy to open with the bounce method. Make a few attempts before you try to pick the lock. If four or five tries do not open the lock, however, it is probably better to stop wasting time bouncing it, as it will not open in this way.

It should be remembered that some types of high-security locks are much more difficult to pick or impression than ordinary ones. The Medeco locks, for instance, are very pick-resistant, as they are based on a dual-locking principle. The rotation of the plug in such a lock is blocked by the secondary locking action of a sidebar that protrudes into the cylinder case. The pins have a slot along one side, and they must be rotated so that this slot aligns with the legs of the sidebar. The tips of the bottom pins are chisel-pointed, and they are rotated by the action of the tumbler spring seating them on the corresponding angle cuts on the key. The pin tumblers must therefore be elevated to the shear line and rotated to the correct angle simultaneously, to allow the sidebar's legs to push into the pins before the plug will turn within the cylinder case. Picking such a lock successfully in the field is generally not likely, though it might conceivably go well under laboratory conditions.

Medeco cylinders and other cylinders of similar resistance are also protected from physical attacks, including wrenching and drilling, by hardened, drill-resistant steel inserts in the lock. Two hardened, crescent-shaped plates within the cylinder case protect the shear line

Figure 16. Protective inserts on the face of high-security cylinders.

and the sidebar, while hardened rods within the face of the plug and a ball bearing in front of the sidebar protect these areas (fig. 16). These inserts are fairly good for protecting the lock cylinder against drilling, but of course the door, other parts of the lock set, or even the wall, might still be easy to breach by physical attack.

High-security locks are generally more resistant to impressioning than ordinary locks as well. Another feature of many high-security locks is that the factory usually maintains control of the key system. The owner must present an I.D. card and sign a special order form to obtain extra keys. These keys, often known as "registered" keys, cannot be manufactured without special equipment, usually available only in a price range prohibitive to ordinary locksmiths. The key can therefore only be copied by the lock manufacturer or its authorized affiliates.

Tubular Cylinder Locks

The tubular cylinder lock is a variation of the standard pin tumbler cylinder lock. As the latter locks became extremely popular, that popularity very quickly extended to the tubular cylinder lock as well. Today these locks are generally relied upon as high-security locks, and as they are very dependable for their price range, they can be found anywhere in many roles, such as key-in-knob locks and desk locks. Yet other common variations are the cylinder locks used on coin boxes, many coin-operated washing machines, and numerous modern vending machines. Larger versions are used as protection on automatic teller machines and in some banks.

One other important application for the tubular cylinder locks is alarm systems. They might be used to protect the control unit or act as a key switch to the entire system. The tubular cylinder lock is a real pin tumbler lock that basically works like any ordinary pin tumbler cylinder lock. The lock and its key are tubular, however, and the pin tumblers are arranged in a circle (fig. 17). In this arrangement, all of the pins can be seen from the outside.

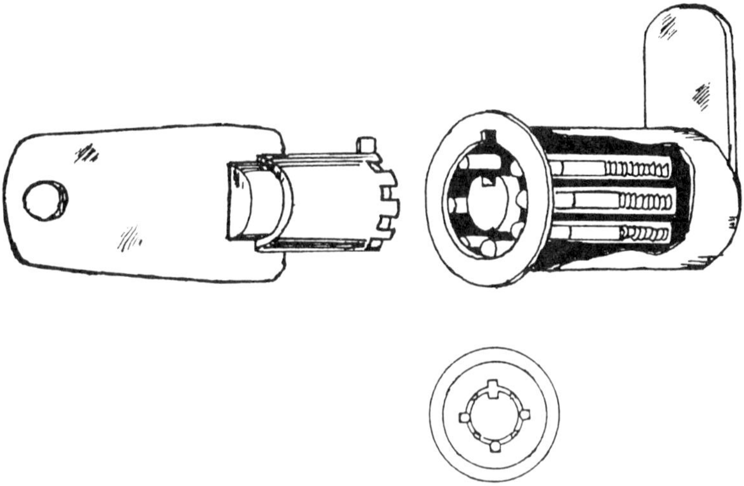

Figure 17. Tubular cylinder lock.

As is the case with the standard pin tumbler lock, the plug will rotate to operate the cam when all the seven or eight pins in the plug have been positioned at the shear line. A tubular key has a hollow, cylindrical-shaped blade that has indentations around its rim. There are usually seven or eight indentations, corresponding to the number of pins in the lock. Exactly as in an ordinary pin tumbler lock, the pin tumblers will be pressed in position by the cut of the key.

A tubular cylinder lock can be picked with a straight pin and a thin but square-shaped torque wrench (fig. 18). It is not easy, however, as you will generally have to pick it several times in order to accomplish the unlocking radius of 120 to 180 degrees. Another problem is that the cylinder will lock after it has been turned slightly. Furthermore, if the lock is left only partially picked, the key will not be able to open it unless you pick it back to the locked position. This will usually take a considerable amount of time. Another option is to use a special tubular lock-picking tool, such as is used by regular locksmiths (fig. 19). Many locksmiths will not bother with this, however, and instead simply drill out the lock if the key is lost.

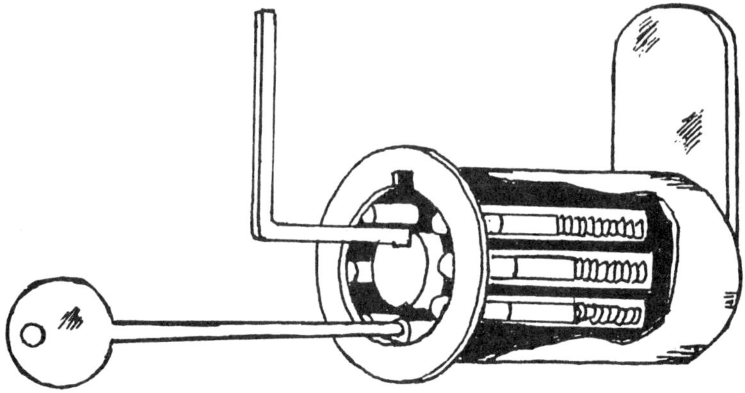

Figure 18. Picking the tubular cylinder lock.

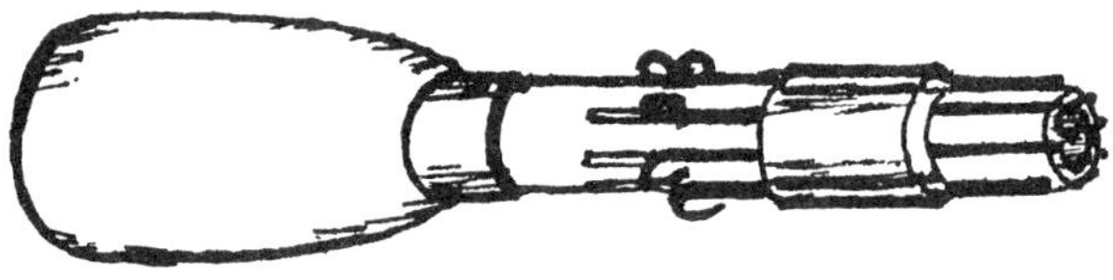

Figure 19. Tubular lock-picking tool.

The tubular lock-picking tool is a hybrid that picks and acts as a torque wrench at the same time. Actually, it does not pick as much as it impressions the lock. The tool has seven or eight (depending on the type of lock) steel fingers that adjust themselves to correspond to the cut depth of the original key. These fingers are held in place by a rubber sleeve or a strong rubber band. The rubber band will be tightened, or another rubber band will be added, once the lock opens, so that the steel fingers will remain in the correct position. At this point, the tool can either be used as a key to open the lock or as a pattern to cut a permanent, regular key.

Vehicle Locks

Vehicle locks can be of almost any type, including the previously described disc tumbler and pin tumbler locks. However, it is also quite common that a vehicle lock instead relies on the sidebar principle. The sidebar lock is a more

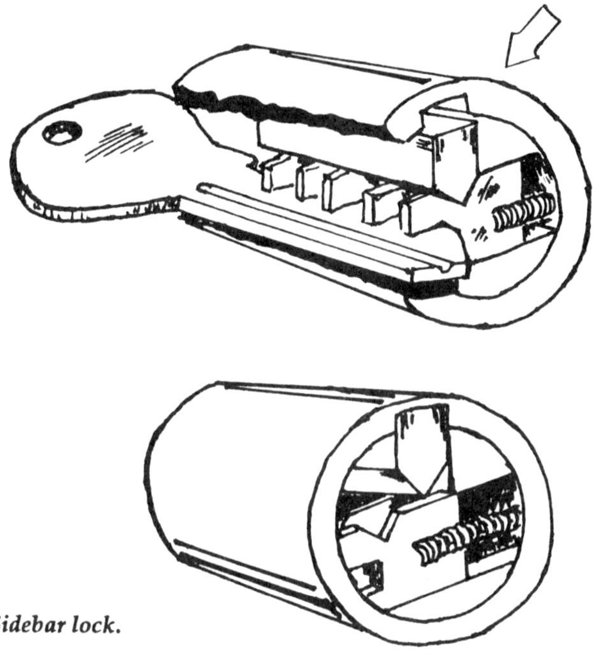

Figure 20. Sidebar lock.

specialized vehicle lock. In fact, it is a variety of the disc tumbler lock. Sidebar locks are commonly used for ignition, door, and trunk locks. They are fairly simple in construction (fig. 20). There are disc tumblers inside the lock with V-shaped notches in their sides. When the right key is inserted and engages the tumblers, the key aligns them so that the spring-loaded sidebar moves out of the cylinder and into the plug. The plug is free to rotate when the sidebar has passed the shear line, which will unlock the lock.

Although the construction of the sidebar lock is simple, the sidebar makes the lock difficult and time-consuming to pick. There is no way to determine when the sidebar will fall in place, as it is impossible to hear or feel the tumblers align with the shear line. Here experience is necessary, unless you want to drill the lock open. Then an L-shaped wire can be used to put pressure on the sidebar while you rake the disc tumblers into place. Such a method will, of course, leave very clear marks, proving that the lock has been tampered with.

The sidebar principle is also used on certain pin tumbler cylinders. Sidebars will only be found in high-security locks, described in Chapter 1.

MAGNETIC LOCKS

Magnetic locks work on the principle that identical magnetic polarities repel each other. In a magnetic lock, there will be a number of small magnets arranged in a certain order. The key contains the same number of magnets. However, these magnets are arranged so as to repel the magnets in the lock. The polarities will therefore be arranged in the same way both on the key and inside the lock. When the magnets inside the lock are repelled, a spring-loaded bolt will be moved to open the lock (fig. 21).

It is impossible to pick a magnetic lock. In an emergency, however, the lock can be breached by exposing it to a suffi-

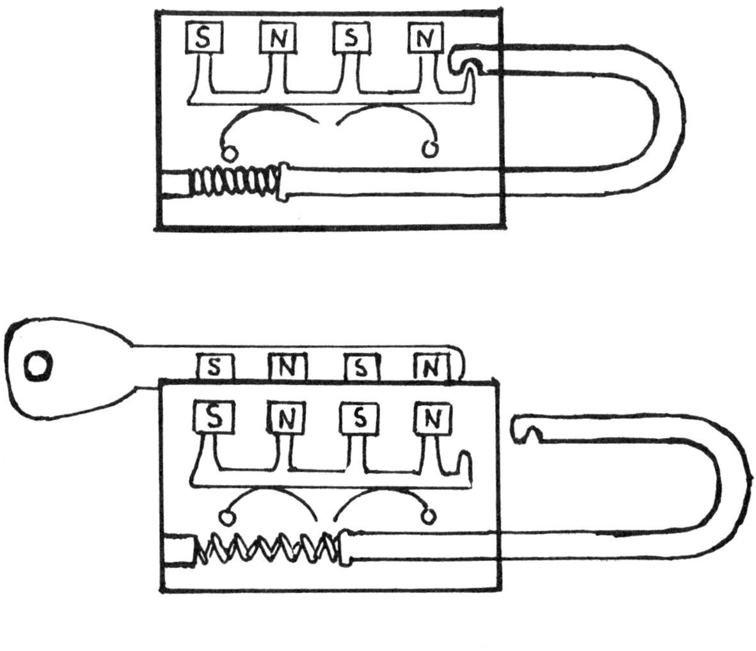

Figure 21. Magnetic padlock.

ciently strong, pulsating electromagnetic field. If the shackle (most magnetic locks seem to be padlocks) or the bolt is pulled repeatedly, the lock will then spring open. However, the electromagnetic field is likely to change the magnetic properties of the magnets in the lock permanently. If this happens, the real key cannot open the lock at a later time. The intrusion will thus be detected easily. The electronic field can be created by a portable field instrument. Of course, a sufficient power source is required.

It should be noted that magnetic door locks are also used sometimes. They are normally opened by a metal or, more commonly, plastic card containing a magnetic strip that has been coded with a certain magnetic combination. The internal mechanism is the same, however.

Simple Suitcase Locks

Almost all suitcase locks are of the simple, warded type. They have only a primitive bolt mechanism to keep the case closed. A very few suitcase locks rely instead on a lever-type mechanism. The lever suitcase locks are usually recognized because of the fact that the key will go half a centimeter or deeper into the lock before turning. A warded lock will be much shallower.

It is easy to make a skeleton key that will open most types of warded suitcase locks. Almost any suitcase key can be used for this purpose. Alternatively, simple suitcase locks can be picked with a special lock pick (fig. 22), easily manufactured from a strip of steel. The lock pick is merely inserted into the keyway, and when the bolt is located, the pick is turned to manipulate the bolt. This will open the suitcase.

Safes and Combination Locks

A combination lock is one that may or may not be operated with a key but can always be operated by entering a combination of numbers or other symbols. This is done either by rotating a dial or pushing buttons.

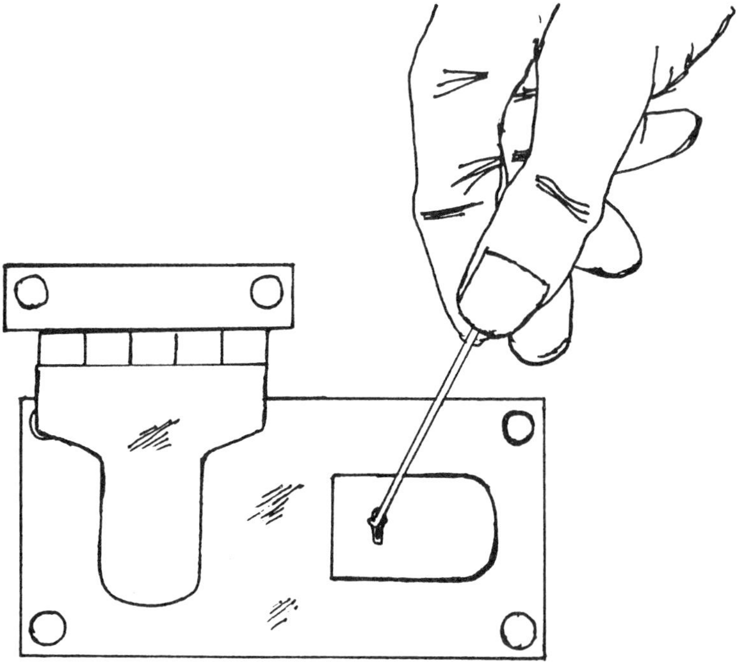

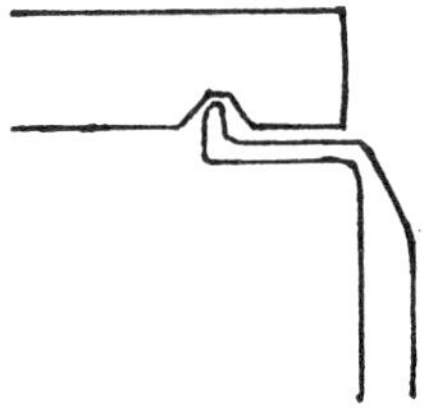

Figure 22. Picking a suitcase lock.

The most well-known combination lock is the safe combination lock. Such a lock consists of a series of interconnecting wheels that rotate around a central core. The device is controlled in its revolutions by an outside combination dial.

All combination locks of the safe type operate on the same principle, even though there are internal differences between different types of locks. An internal wheel pack will be rotated by manually rotating the external combination dial. The wheel pack consists of a series—usually three but

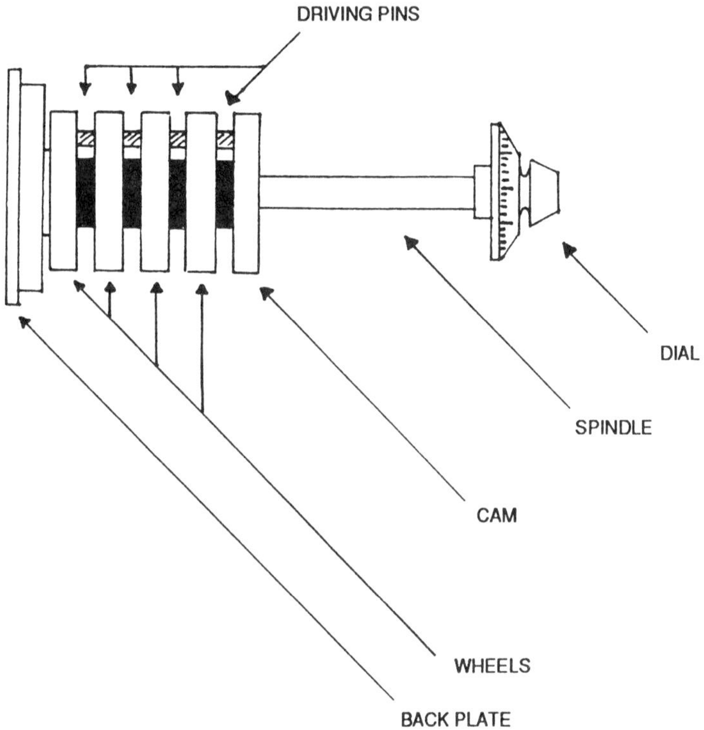

Figure 23. Dial combination lock.

sometimes four—of interconnecting wheels, or tumblers. As the dial has one hundred different numbers, or positions, a three-wheel mechanism will have almost a million possible combinations. The four-wheel combination locks have almost 100 million possible combinations. Each wheel is designed to align its gate with the bolt-release mechanism only after a certain number of revolutions and a certain degree of rotation. This design can be programmed quite easily so that the combination that will open the lock can be changed according to the owner's wishes.

Basically, a combination lock of this type works in the following way. In addition to the wheels, the complete mechanism also consists of a dial, a spindle, and a driving cam (fig. 23). These devices form the driving mechanism that moves the wheels into locked or unlocked position.

The wheels are moved by driving pins that are affixed to the back of the driving cam as well as to the wheels. These driving pins will engage and disengage the wheels as the cam revolves according to the movement of the dial. When the dial (by way of the cam) has set all wheels in the right position and is revolved slowly either back in the opposite direction to the last combination number or, in newer safes, revolved back and then forward once again, it will operate the bolt of the lock to the open, unlocked position.

The wheels are always rotated in order, and the number of turns depend on the number of wheels. The most common lock type, the one with three wheels, requires a 3-2-1 rotation sequence, while most high-security combination locks have four wheels and require a 4-3-2-1 sequence. The combination dial is always rotated (in the basic model) first three turns, then two turns in the reverse direction, and finally, in the opposite direction again one turn. Most four-wheel combination dials are designed to begin rotating to the left, but this is not universally true. Likewise, most three-wheel combination locks are designed to begin rotating to the right. There are, of course, differences between different types of safes. As the wheels are rotated, the gates will be aligned by stops, one for each wheel and one on the wheel-pack mounting plate. The bolt will be free to release only when all gates are aligned.

Older and inferior types of combination locks can be distinguished by audible clicks when the wheels rotate. This allows a skilled individual to manipulate the lock without knowing the combination in advance. Contemporary combination locks have at least three false gates in every wheel (fig. 24), so that manipulation will be much more difficult. It can still be done, but only by an expert who has had lots of practice as well as special training and knows the peculiarities of the particular type of lock he is working on.

Manipulating a combination lock in this way is a matter of touch as well as hearing. The latter is usually assisted by using an electronic stethoscope, but the sense of touch can only be developed by long training. The process can be de-

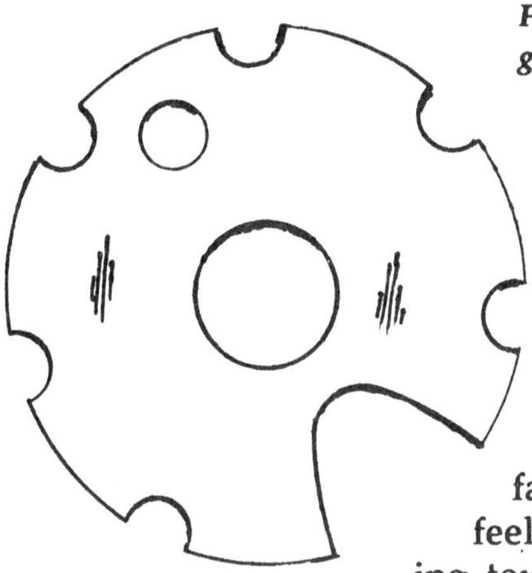

Figure 24. Wheel with six false gates.

scribed in the following way, although the inexperienced should not attempt it during a real operation.

Rotate the combination dial slowly until you hear a very faint click. You will then feel that the bolt is hesitating, touching the *far side* of the gate. At this point, move back one number on the dial and note this number. This is the first number of the combination (unless it happens to be a false gate, of course). Then turn the dial in the reverse direction and proceed slowly until you have passed the first number two or three times, depending on the type of lock you are working on. As you continue to turn the dial, you will notice that the bolt touches the far side of another gate. Once again, note the number preceding the one at which this occurs. This is the second number of the combination. Then turn the dial in the reverse direction—the original direction—once again until the process repeats itself.

After following this procedure, if you have not been tricked by any false gates, you will have the correct numbers of the combination. However, it is by no means certain that you will have them in the right order. This is no great problem, though, as you can determine the right sequence simply by varying the sequence of the numbers until you hit the right combination and the lock opens.

The existence of several false gates will naturally delay this process considerably, but the working principle remains the same. Remember, though, that practice and more practice is the only key to success. No type of lock picking can be learned properly in a short time, and this is especially true for combination lock manipulation.

Another important thing to remember is that there will always be at least a certain number of divisions on the dial between the different numbers of a genuine combination. This is to avoid any possible malfunctioning within the combination lock mechanism. This means that if you have determined two numbers that are too close to each other, or to zero, then one of them will almost certainly be false.

One important point that will assist the manipulator should be mentioned here. It is very common that the owner of the safe does not realize that the dial must be revolved completely a number of times (four times for a three-wheel lock and five times for a four-wheel lock) in order to lock the safe properly. If the dial is only partially rotated, the lock will not be locked properly, as only the last wheel will have been disengaged from the unlocked position. The remaining wheels will, of course, remain in the unlocked position. Naturally, this will greatly benefit the intruder who tries to manipulate the lock open.

Of course, the intruder will not know if it is locked properly or not. He will also be unaware of whether the dial, if the lock was improperly locked, was last turned to the left or to the right. To find out these important details, he must first of all turn the dial very slowly in either direction, simply to feel whether the driving cam is engaging the wheel or not. If it is, he must reverse the direction of the dial at once without going further, or he will lock the lock. If he can turn the dial in the opposite direction without the driving cam engaging the wheel, the lock is not locked correctly.

When it is in this way determined that the lock is improperly locked, the operative will move the dial to the position which he knows from previous study is the one in which the gate of the cam will be aligned opposite to the fence. This is generally position five or ninety-five, but it varies in different types of safes. By depressing the dial, and possibly moving it slightly to the left or right, he now attempts to cause the wheel to align itself in order to open the lock. If the wheel is only slightly disengaged from the unlocked position, and if the operative is lucky, this simple procedure might open the lock.

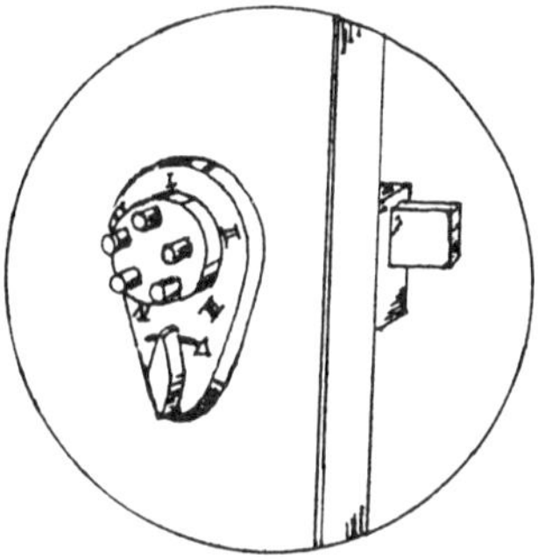

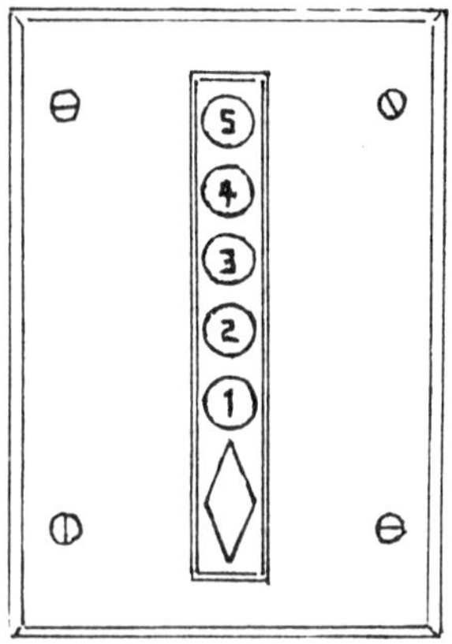

Figure 25. Push-button combination locks (top) and a digital keypad lock (at right).

Unfortunately, it is much more common that the last wheel of an improperly locked combination lock is completely disengaged. Then the intruder must concentrate instead on finding the correct number on the dial that will bring this wheel to the unlocked position. This can be done in the regular way described above. Only one number need then be found in order to unlock the safe.

Yet another point to consider is that somebody who is too careless to lock his expensive safe properly will probably use an easily remembered combination. Certain numbers, such as his birthday, might have some meaning for him, or he might have used numbers divisible by five or ten. Combination lock manufacturers advise against such combinations, but because easy combinations are easy to remember, careless individuals still use them. It might also be worthwhile to search the room for a written note of the combination. Often such a note can be found in a desk drawer or in a notebook on a nearby table.

Some combination locks are interlocked with timer locks. This will effectively keep the lock closed until a certain time. Such advanced locks are generally only found in bank vaults, however, so they rarely present a problem.

Nowadays push-button combination locks (fig. 25) and digital keypad locks are also in widespread use. The digital lock is an electronic lock that can only be opened by keying in the appropriate number code. The push-button combination locks usually consist seemingly of only a control knob and a number of push buttons to control the lock. They are sometimes electronic, but not always. Mechanical versions are also common. Nevertheless, these locks are classed as combination locks, as they have no key. Push-button combination locks are now becoming more and more common in hotels, motels, and government institutions, for instance. Many companies also use them.

In companies and institutions, some high-risk areas can be protected by two locks, an ordinary one locked after office hours only and a keyless push-button lock allowing access by the staff during the daytime. Hotels and motels frequently find that combination locks of these types save time and money, so they are becoming more popular each day. The combination can be changed every time a guest has checked out. Therefore, the level of security is high.

Digital keypads are often used at the common entrance to an apartment or office building. Unlike some push-button types, these locks are always electronic. The lock will be opened when somebody punches the correct number code.

The standard push-button combination lock has five push buttons. The combination, pressed in the right order, will allow the knob to be moved and the lock opened from the outside. This lock can be opened with the combination only.

Another model is the key bypass lock. This lock can be opened with either the combination or the key. Employees or tenants use the combination, while senior management personnel will use master keys. This lock can be picked. A lock of this type has no special advantages, so they might be less common in the future.

Most, but not all, of these locks will include an automatic spring latch that locks the door when it is closed. This is to ensure that nobody forgets to lock the door again after entering. Another common feature is a face plate shield to prevent anyone from observing the push button operation from a distance in order to learn the combination. The face plate is the visible part of a lock and therefore the weak link if somebody is observing the area. Yet another option is to eliminate the latch hold-back feature so that the lock can never be kept open. It will then remain locked at all times, except for a brief moment when the correct combination is used.

Certain locks will also allow two push buttons to be pushed at the same time, in effect producing a different number, in order to raise the security level. However, these locking units cannot use the same push button more than once in a given combination.

The majority of the push-button locks work in the following way. First turn the control knob to the left to activate the push buttons. Then press the push buttons in the correct combination. Finally, release the last push button or push buttons before turning the control knob to the right. This will open the lock. The lock can then be relocked by turning the control knob to the left, or, alternatively, relocking will be an automatic feature.

The lock cannot be opened by removing the control knob, as this knob is connected to the lock by a friction clutch. The internal mechanism of the lock will therefore be damaged if the control knob is forced or removed.

Other push-button combination locks employ four-digit or seven-digit combinations. In these locks, there are usually ten push buttons to choose between. These units can frequently be unlocked by either a four-digit change combination or a six-digit master combination. The master combination might work for several different locks in exactly the same way as a master key. Although these locks may be electronic, they are often mechanical. We will deal with electronic locks in more detail in Chapter 6.

Sesame locks (fig. 26) are extremely simple combination locks that are frequently found on briefcases and suitcases. These consist of three dials numbered from zero to nine. The number of possible combinations is very low—only one thousand. This means that the correct combination can

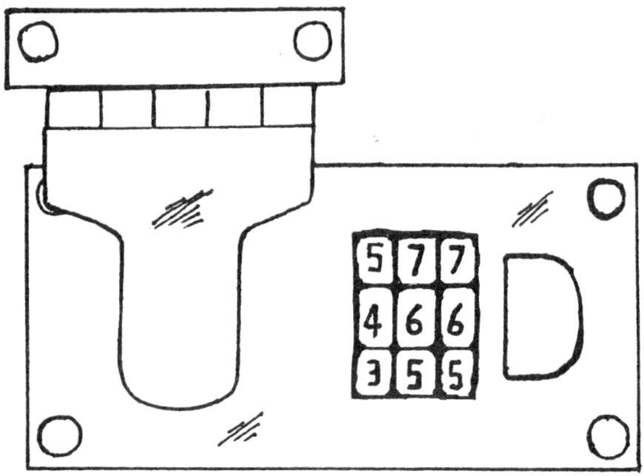

Figure 26. Sesame combination lock.

be found by trial and error if sufficient time is available. Expect to spend about half an hour to go through all possible combinations. Less time is usually enough, as the correct combination is often found long before you have checked all of them.

Begin the process by setting all three dials to zero. Then move dial C to 1, 2, 3, 4, 5, 6, 7, 8, and 9, one after the other, while pressing the catch after each change in number. Unless the correct combination has been found, change the setting of dial B to 1. Dial A will remain on zero. Once again, repeat all numbers on dial C while pressing the catch as before. If the correct combination has not been found, set dial B to 2, and repeat the process. Continue the process until dial B has been set to all possible numbers. Then set dial A to 1 and repeat the entire operation once again. Dial A is then set to 2, and the process is repeated.

This process, although boring, is in fact easy to perform and will sooner or later result in the lock being opened. Do not try to think of what you are doing; simply learn the process by rote and do it.

Combination padlocks with a dial can be opened in basically the same way as other dial-type combination locks, but they are usually simpler in construction and are consequently easier to open. For this reason, they are often used for practice. Remember, though, that it is easier to locate the gates within the combination padlock if you pull out the shackle as you rotate the wheels. It is also easier to open combination padlocks that have been in use for a long time because the gates on the wheels will have become smoothed down, which will simplify the manipulation of the wheels.

Yet another type of combination padlock is the sesame padlock (fig. 27). This padlock has no dial but works instead on the same principle as the sesame lock described above. However, the sesame padlock has four combination wheels, numbered from zero to nine. The method used to open a sesame lock with only three wheels is therefore not very practical on this padlock, as the required time to check all possible combinations would be approximately ten times as long, or five hours. There is, however, another way to open this lock.

The four-wheel sesame padlock is designed to unlock the shackle only when each wheel is positioned so that a flat spot on each wheel is aligned with the corresponding flat spots on the other wheels. Each wheel

Figure 27. Sesame padlock.

Figure 28. Manipulating the sesame padlock.

has a changeable hub with such a flat spot. These flat spots, when aligned toward the side of the lock stamped with the trademark, will unlock the padlock. The current combination of the padlock determines where the flat spots will be located on the wheels.

A special tool (fig. 28) made of very thin steel can be used to locate these flat spots. Insert the tool into the lock between the wheel and the housing. Turn the wheel slowly and try to locate the flat spot. When it is found, either add or subtract five from the indicated number. This will give the correct combination number of that wheel.

Another, not very obvious type of combination lock is the remote-control lock. Such a lock is activated by an infrared beam from a hand-held device. Remote-control locks are generally used only on garage doors and driveway gates. Although the signal opening the lock is supposed to be unique, these locks are not very secure. This makes no difference, as these devices are never used to protect really important positions. See Chapter 6 for more information on remote-control locks.

Padlocks

The padlock was first invented by the ancient Romans. Today a variety are in common use, including warded, lever, disc tumbler, pin tumbler, and combination types.

Padlocks can usually be picked quite easily. Hold the padlock with the same hand with which you are using the torque wrench. It is generally easiest to do this by holding the padlock between the thumb and the forefinger. Then you can hold the wrench with your ring and little finger.

Your main hand is then free to work the lock pick. Take some time practicing how to hold the padlock, though, and make certain that it is easy to work on it. Different people sometimes prefer to hold the padlock in different ways.

Padlocks are not really that different from standard locks of the same type. Some smaller and simpler warded padlocks, for instance, have a single ward only and take very simple keys. Otherwise they are designed in the same way as other warded locks. The majority of the warded padlocks have three wards, or at least two. The key must pass through the wards before it can disengage the spring bar from the slot in the shackle end.

A warded padlock can be defeated by a skeleton key.

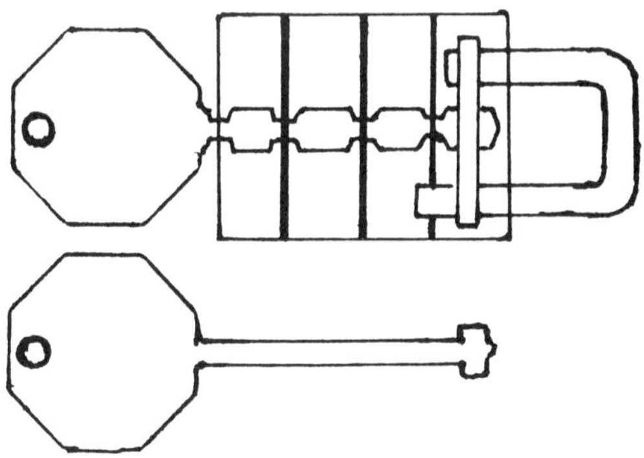

Figure 29. Warded padlock with the internal wards illustrated and a suitable skeleton key.

This is an ordinary key, but it is ground down (fig. 29). Note that skeleton keys are illegal in many locations. There are a few basic shapes, one of which will almost certainly defeat the lock. See the first section of this chapter.

Most warded padlocks can also be picked with an improvised T-shaped wire pick (fig. 30). Such a pick is easily improvised from a piece of stiff piano wire. An old method of picking padlocks is to insert a hat pin through

Figure 30. T-shaped wire pick.

the keyway and then use it to disengage the shackle bolt. This still works with older types of padlocks, but not with the recent, more secure ones. Warded padlocks can also be impressioned easily. The procedure is the same as with ordinary warded locks.

The other types of padlocks, including the cylinder types, can be picked or impressioned in roughly the same way as the ordinary locks of their type. Disc tumbler padlocks can also be picked, but it takes practice. Another possibility is sometimes, but not always, to acquire a set of test keys from a locksmith supply house. Such keys will facilitate the work.

Whenever picking a padlock, remember that in some padlocks you need to pull the shackle in order to help release it from the locking spring. If this does not help, repeatedly work the shackle in and out while picking the lock. This will unlock the mechanism eventually.

Master Key Systems

Master keys can be made because a single key can be cut to match several lock combinations. Master keying always relies on coding systems that allow the locksmith to distinguish various key cuts and tumbler arrangements. In most key coding systems, the tumblers can be set to any of five possible depths. Since most locks have five tumblers and each one has five possible depth settings, there can be thousands of different combinations.

Warded locks and lever locks may also be master-keyed, but the security level will then be low. Disc tumbler locks, and especially pin tumbler locks, are more commonly used for this purpose.

Disc tumbler locks, when adapted to master-keying, are peculiar in that the master key will use a completely different keyway, located next to the regular one. This can be seen by closely inspecting the lock. The master key will operate on the left side of the tumbler, while the change key, the regular key, will operate on the right side (fig. 31).

Pin tumbler locks are master-keyed by adding another

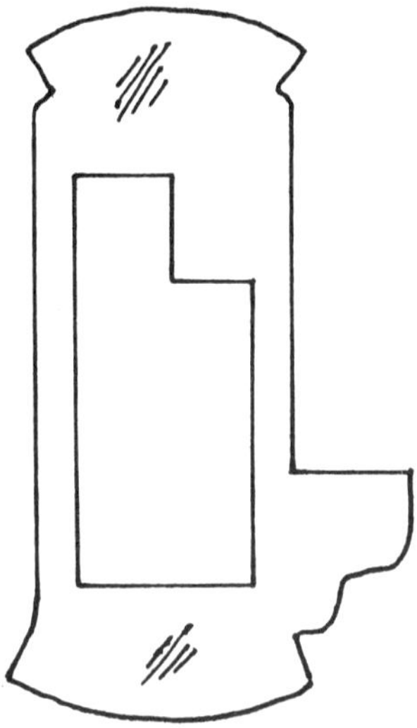

pin, the master pin, between the top and the bottom pin in at least one pin chamber. The master key will have some cuts identical to the change keys, but it will operate the lock by raising the other pin or pins to the "new" shear line created by the master pin (fig. 32). What actually happens is that there will be two breaks for the shear line. In effect, this makes the lock slightly easier to pick, as

Figure 31. Disc tumbler prepared for master keying.

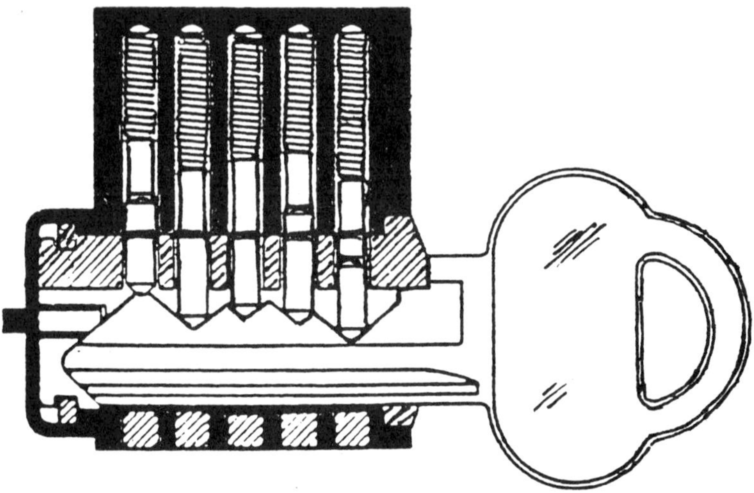

Figure 32. The master key relies on raising one or more master pin tumblers to a "new" shear line.

there are more possible combinations that will raise the pins to the shear line. In master-key systems with many different locks, there might be one or more such master pins between all regular pins. In this case, the system allows for numerous locks, and also several submaster keys. These systems are usually factory designed and manufactured and can include as many as four breaks at the shear line.

A master key will not open all locks in a building, unless it is a so-called building master key or an emergency master key, which is a top-level master key that will operate all the locks at all times. It is more common to encounter, for instance, an engineer's key, which is a selective master key that is used by various maintenance personnel. It is therefore necessary to ascertain the level of the master key copied or otherwise obtained before it is used in an actual entry operation. In some low-security systems, the level might be stamped onto the key, although today it is more common to keep a master-key system chart in a safe place, detailing all the various key numbers and the key's position in the key hierarchy. The key will then only be identified by a code.

A master-key system is always divided into several different hierarchies, or key levels (fig. 33). If a master key can be obtained, it is advisable to identify its place within the hierarchy, as this will indicate which locks you can use the key in.

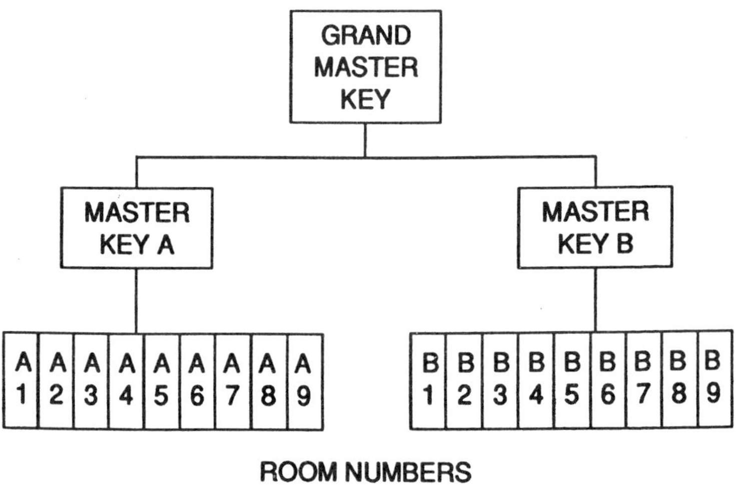

Figure 33. An example of a master key hierarchy.

CHAPTER 3

Improvising Lock-Picking Tools

Picking a lock is not easy, and, with the exception of the simplest locks, it takes a considerable amount of time. It is therefore usually better to rely on an experienced locksmith to make a new key to the lock, especially if you have to enter the area protected by the lock more than once. He will then utilize the various impressioning techniques described in the previous chapter. However, in emergencies it might still be necessary to pick a lock.

A definite prerequisite for lock picking is a very good knowledge of the lock and how its mechanism works. It is also usually necessary to have the proper tools—lock picks—to do the job. Actually, a pick is not usually a special tool, but rather any device that is used to manipulate the tumblers in a cylinder into an unlocked position or to bypass whatever device is protecting the lock from being opened. Lock picks can therefore be improvised easily, as long as some basic materials and a few simple tools are available. Remember, though, that without a proper locksmith's license, the possession of lock picks is illegal in many locations.

It is not really necessary to carry lock picks on one's person except when they are definitely required. Lock picks can be made from easily obtained raw material almost anywhere, and the only essential tools are a pair of pliers and a

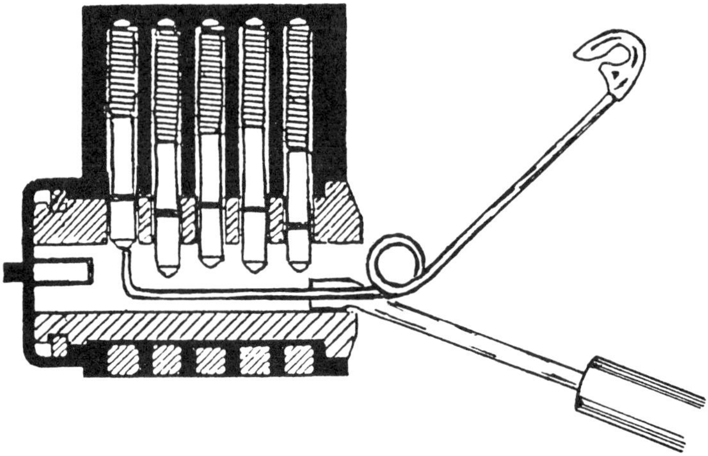

Figure 34. Picking a pin tumbler cylinder lock with a safety pin and a small screwdriver.

file. These tools, unlike the actual lock picks, do not gain the attention of the police or enemy if an operative who is in possession of them is searched.

In a real emergency, whatever is at hand can be used as a lock pick, although the work will then often be slightly more difficult. It is, for instance, quite possible to use a large safety pin and a small, slightly bent screwdriver instead of a pin tumbler lock pick and a torque wrench. Simply bend the tip of the safety pin at a 45-degree angle so that you can use it inside the lock (fig. 34). If no pliers are available, it is easy to bend the safety pin with the help of the keyway of the lock you are going to pick. The safety pin should be quite large, at least four centimeters long, so that you will be able to hold it easily while you are working on the lock. It helps if the tip is filed flat, as this will make it easier to locate and raise the pins.

"Real" lock picks come in numerous shapes, depending on the type of lock to be picked (fig. 35). Lock picks can be made easily from flat, cold-rolled steel that is less than half a millimeter to a little less than a millimeter thick. The actual thickness is really not very important, except that the pick must be able to enter the lock and work in there. The

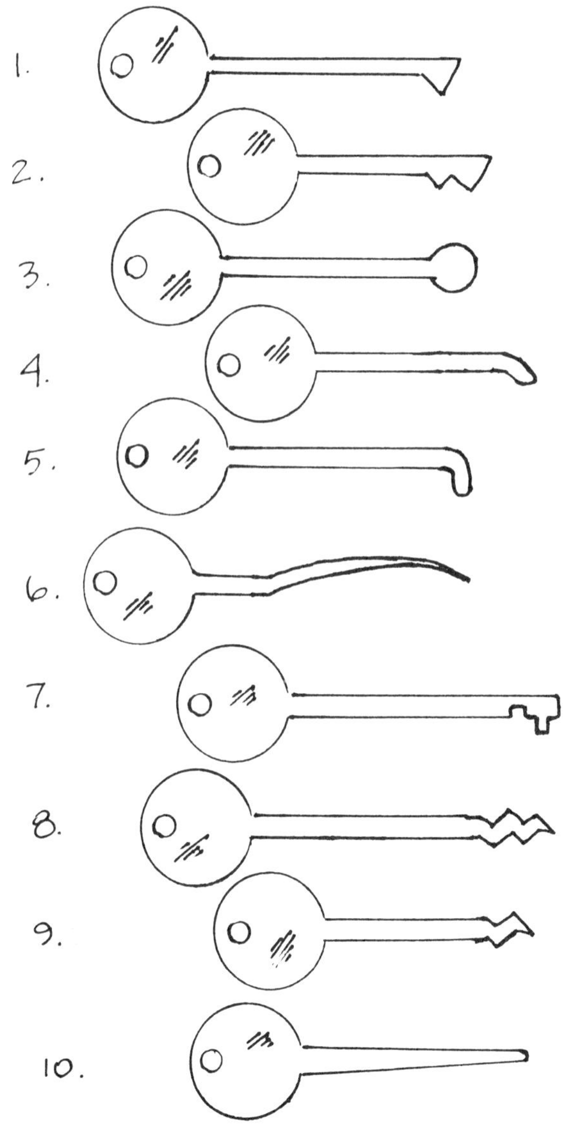

Figure 35. Various types of lock picks: 1. diamond lock pick, 2. double diamond lock pick, 3. circular or ball lock pick, 4. half-round feeler lock pick, 5. round feeler lock pick, 6. reading tool, 7. flat lever lock pick, 8. rake, 9. half-rake, 10. tubular cylinder lock pick.

steel strip should be about 15 centimeters long and at least 8 to 10 millimeters wide. Some people prefer to have one end fitted with a handle, while others prefer their lock picks to carry working surfaces on both ends.

Make your lock picks by grinding down the steel strips to the correct size with a file (the best option) or a grinder and a carborundum wheel. If the latter is used, take care that the metal does not become too brittle because of excessive heat.

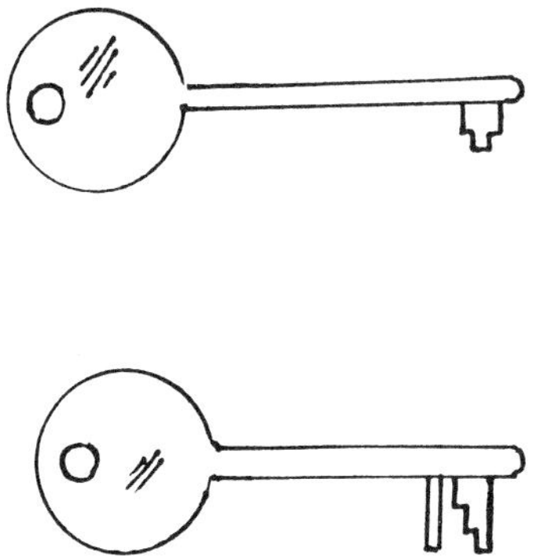

Figure 36. Lock picks for warded locks.

As was mentioned in Chapter 2, you will need lock picks of slightly different design for warded locks (fig. 36). These picks are easy to make. In many instances an old precut key, ground down to pass the wards in the keyholes, is quite sufficient. Then it will in effect become a skeleton key.

One of the most useful lock picks is the diamond pick (see fig. 35). It can be used for most pin and disc tumbler locks. Most experienced locksmiths seem to use this pick much more than any other. It is really an all-around lock pick.

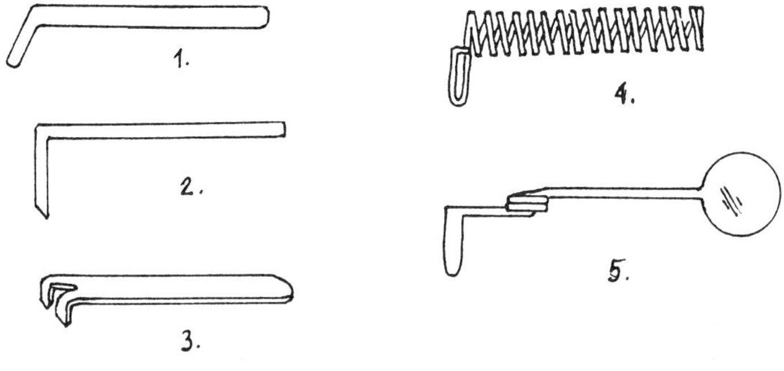

Figure 37. Torque wrenches: 1. & 2. basic torque wrenches, 3. torque wrench for double-sided disc tumbler locks, 4. & 5. steel spring torque wrenches.

Usually some kind of torque wrench must be used with the lock pick. These tools are made of hardened spring steel, around 7.5 to 12.5 centimeters in length and a little more than a millimeter thick. Sometimes longer torque wrenches are required, as these tools must be long enough to reach the interior of the lock. There are numerous variations, but the two most useful ones are the basic torque wrench and, for the really delicate work, the steel spring wrench (fig. 37).

Certain locks (double-sided disc tumbler locks, for instance) require special torque wrenches. Such picks and wrenches are more difficult to improvise (fig. 38) but can be acquired from a locksmith supply house.

The L-shaped lever tumbler lock pick can be made from spring steel no more than 2 to 3 millimeters thick. A torque wrench of the same thickness is also useful (fig. 39). Some individuals use a rake pick. As

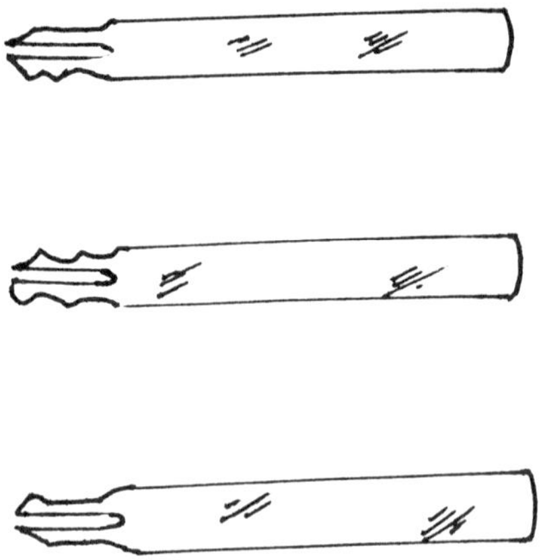

Figure 38. Lock picks for double-sided disc tumbler locks.

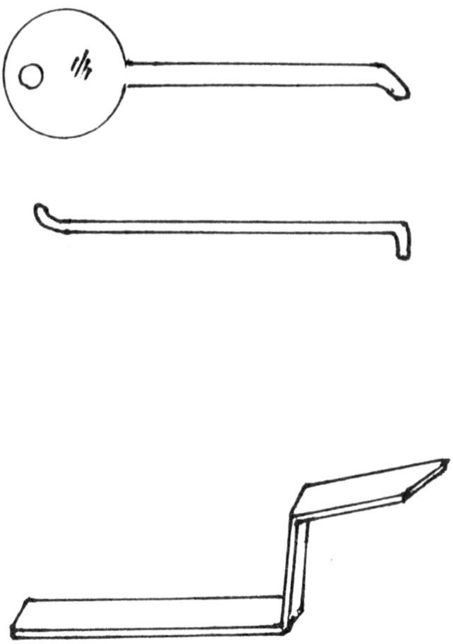

Figure 39. L-shaped lever lock pick, the same pick improvised from wire, and a suitable torque wrench.

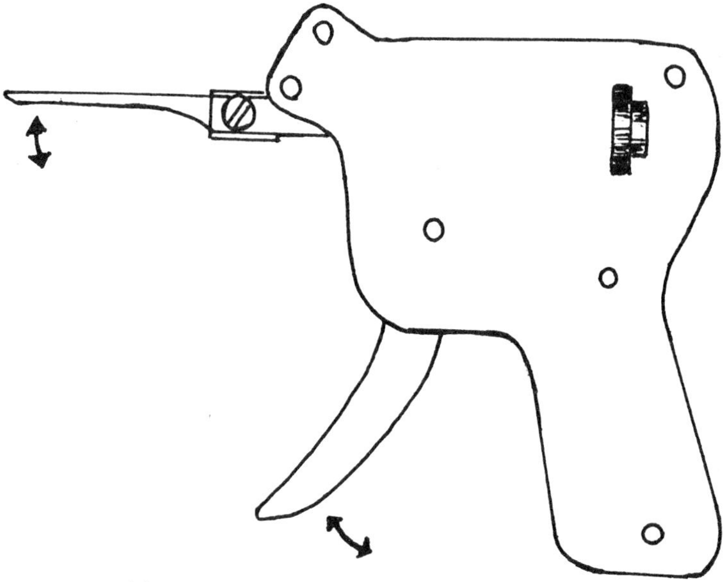

Figure 40. Pick gun.

was previously mentioned, such a pick is not very useful. If a rake pick really is desired, you can just as well use an ordinary diamond pick (see fig. 35) for the same purpose. Although some people make a living from manufacturing and advertising numerous types of lock picks, remember that it is not really necessary to have them all. In most cases, a simple pick will be sufficient. An operative who cannot pick a lock with a diamond pick is usually no better off with any other type of lock pick.

Finally, the device known as a pick gun (fig. 40) should be mentioned. This is a hand-held device that can be used to bounce the driver out of the plug in a pin tumbler lock and into the cylinder case. No great skill is required in order to use a pick gun, but this device is not necessarily faster, nor even better, than picking the lock manually. The pick gun cannot pick all types of locks and is therefore of more limited value than an experienced locksmith. The only time the pick gun might be really useful is when you are picking a lock equipped with mushroom or spool drivers.

The pick gun is inserted into the keyway so that its pick is barely touching all the bottom pins. When the trigger is pressed, the pick gun will rap the pins up to the shear line so that a turning force can allow the plug to rotate and the lock to open.

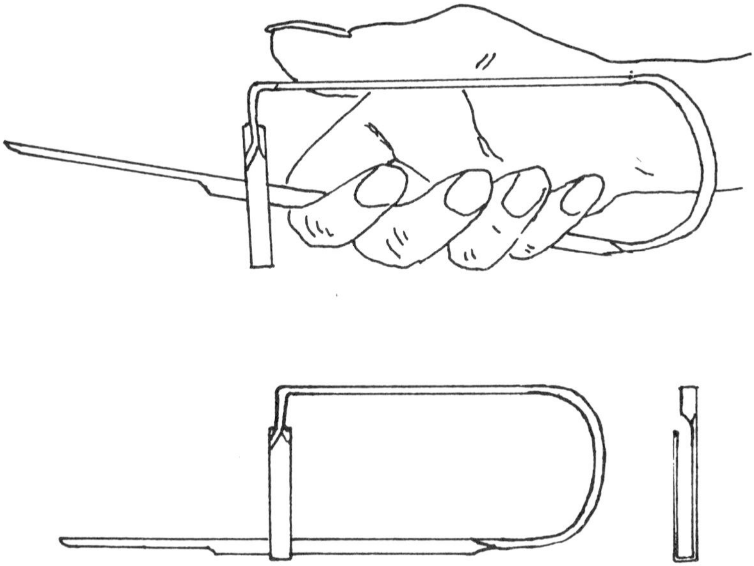

Figure 41. Snap pick.

There is also a manual version of the pick gun. This is known as a snap pick (fig. 41) and can be made easily from spring steel. Just as with the pick gun, insert the pick portion of the device into the keyway and hold so that it is just touching the bottom pins. Then press the upper part down with your thumb before quickly releasing it again. This will rap all the bottom pins. As these pins remain relatively stationary, the force will be transferred to the drivers, which are forced to move upward, compressing the spring. As the area at the shear line will then be open for a split-second, a slight turning force will allow the plug to rotate.

The principle of both the pick gun and the snap pick is

illustrated easily by the fact that force will be transferred through stationary matter. This can be seen in an experiment with three coins on a table. If the coins are lined up touching each other and the coin in the center is held firmly to the table with a finger, the coin on one side will move if the coin on the opposite side is pushed to strike the stationary center coin. The momentary force will be transferred through the stationary coin and instead move the opposite coin. In the same way, the drivers will be moved by the force applied to the bottom pins.

When working with a pick, always remember to use the narrowest pick you have available, as this will give you maximum working space. The pick should be held in about the same manner as a pencil. No wrist action is desired, as only the fingers are dexterous enough to manipulate the pick inside the lock. Wrist action will only make you tired. Lock picking is not a matter of physical strength.

It is also helpful to steady your hand by placing your little finger against the door when you are working on the lock. If the lock is a key-in-knob cylinder, however, steady your hand against the edge of the knob instead.

The pick should be able to enter the keyway above the torque wrench without moving any of the tumblers. If this is impossible, then the torque wrench is either too high in the lock or the keyway grooves are such that the torque wrench must go in at the very top of the keyway. In these cases, picking the lock will be difficult.

Despite the control apparatus established in order to make sure that only professional locksmiths get hold of lock-picking tools, there are various ways of acquiring them. Most manufacturers and distributors of locksmithing equipment refuse to do business with anyone but professional locksmiths. However, they are quite satisfied to send their products by mail to anybody who can prove in any way that he has these qualifications. I say "in any way" because I have found out that a fancy letterhead, business card, or photocopy of a forged locksmith license is usually enough. A photocopy of your advertisement in the Yellow Pages will also be acceptable as proof of professional status. After all, money does not smell!

Many lock-pick sets, both commercially available and improvised ones, are designed to be hidden in various objects so as not to compromise the user. One popular design hides the picks inside a pen cover or in the handle of a hobby knife. In the latter case, the hobby knife can also be used as a handle for the pick, making it easier to hold. A lock pick set might also be hidden inside a jackknife handle or in any other suitable location; the only limit to the possibilities is the operative's imagination.

As a final point, always remember that a real key is quicker and safer than any lock-picking skills. Even if it turns out to be impossible to acquire one of the keys to the building you need to enter, there might be ways of at least inspecting the original key. If a key can be handled for some time, even if it cannot be taken out of the building, it is almost always quite possible to make an impression in wax or a full-scale drawing, and then make a copy at a later time.

Naturally, the full-scale drawing must be done with extreme care, including detailed measurements of every part of the key. This is generally easy, although tedious, except when it comes to the diameter of the key. In certain types of locks, the diameter can often be measured instead by wrapping a paper clip around the original key. The paper clip can then be taken out of the building. In this way, the need for special equipment for measuring the key is avoided.

CHAPTER 4

Other Means of Illegal Entry

Window Entry

There are various ways of opening an unlocked window from the outside, depending on the type of the window. The easiest one to open is the sash window. This consists of two halves, the top and the bottom, that slide up and down. The latch is located between these two halves. One of the oldest

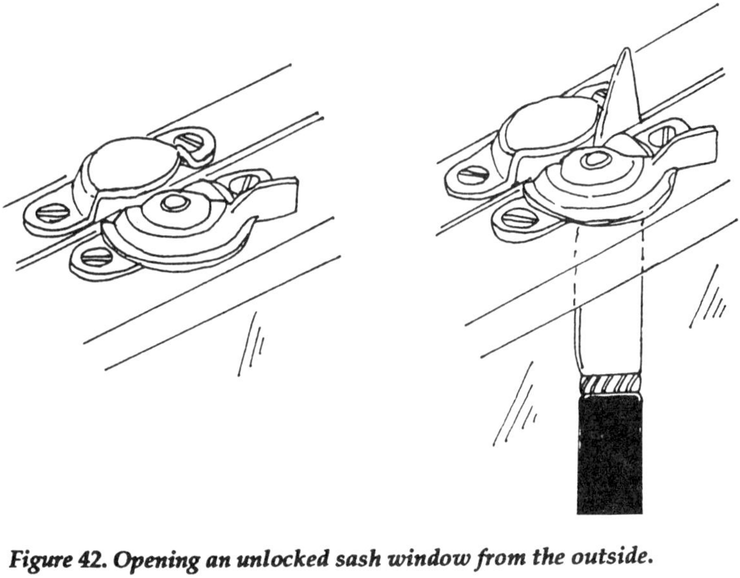

Figure 42. Opening an unlocked sash window from the outside.

tricks of the trade is to slip a knife shim or similar device up between the sashes, or upper and lower halves of the window, to move the window latch to the open position (fig. 42). The window can then be opened easily.

If the area between the window halves is too narrow for a knife, then you can drill a narrow hole at an angle through the wood molding to the base of the latch, insert a stiff wire through the hole, and push the latch back. Hide the hole with paint or dirt afterwards. Unfortunately, contemporary fasteners cannot be manipulated in this way. The fitch fastener, for instance, is a pivoting device with a snail-like cam that cannot be knocked back. The Brighton fastener, another type, relies instead on a screw-down acorn that clamps the sashes together securely. These fasteners also sometimes contain integral locks.

Sometimes the upper and lower window half, or sash, will be connected by means of a bolt slipped into a hole drilled through the lower sash and partially into the upper sash. This was commonly done in older buildings to conserve energy as well as increase the security of the window. If this is the case, simply drill a small hole to the bolt and

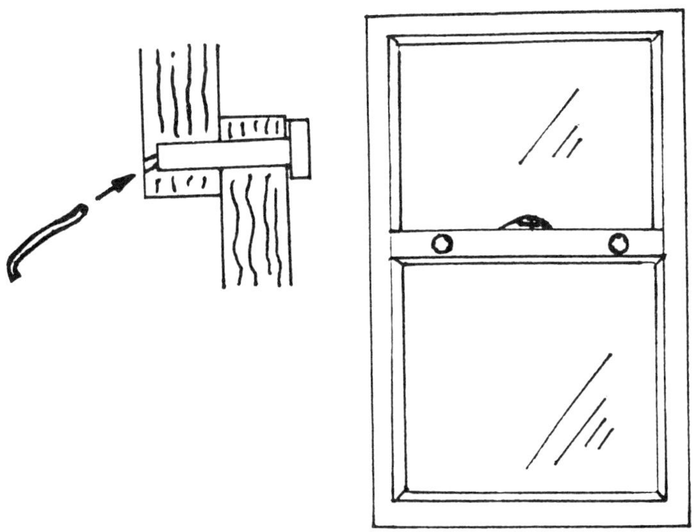

Figure 43. Opening a bolted sash window from the outside.

then push the bolt out of the upper sash with any pointed object, such as a nail (fig. 43). The small hole will usually be difficult to see, and it, too, can be hidden with putty, paint, or dirt.

Other types of windows can also be opened from the outside if a narrow hole is drilled in order to allow a piece of wire to be inserted. Unfortunately, these holes must be drilled in locations where they are much more likely to be noticed by the inhabitants of the house. Another complication is that the windows might be locked. All windows, including sash windows, can be fitted with locks. These come in various types, but all of them are of fairly simple construction. This is of little help, however, as the locks are impossible to open from the outside except by breaking the window.

TRANSOM ENTRY, DOORS, AND DOOR CHAINS

Picking a lock is not the only means of entering through a locked door. There are various other methods, most of which have been in use for quite some time. There are, for instance, ways of entering through a transom and ways of entering by manipulating the lock construction without actually picking the lock.

If a door warbles slightly and the lock is of an older construction that lacks a deadlocking function, the bolt can be retracted with a celluloid strip by the process known as "loiding." This process involves slipping a flat object between the bolt and the strike. The strike, or striking plate, is the part of the locking arrangement that receives the bolt, latch, or fastener. It is recessed in the door frame. A little pressure might allow space enough to insert a celluloid strip between the lock and the striking plate. Then only a slight pressure on the inserted celluloid strip will force the latch back and release the lock. This will open the door.

The door might, however, be fitted with a so-called antipick latch. This is a spring latch fitted with a parallel bar that is depressed by the strike when the door is closed. The depressed bar will prevent the latch from responding to external pressure of any kind. This naturally makes loiding impossible.

A variation of this method is to insert a thin knife, a linoleum one, for instance, between the door and the jamb. The knife's point should be tipped upwards. Then insert a pry bar above the linoleum knife in order to spread the door slightly. Then the latch can be disengaged. Bring the linoleum knife forward, push the latch bolt back into the locking assembly, and open the door. Of course, if there is sufficient space between the door and the frame, the linoleum knife alone is sufficient to move the bolt back.

If, however, there is no space to insert a pry bar, wooden wedges must be used. One wedge should be inserted on each side of the bolt, about 10 to 15 centimeters away from the bolt assembly. This will spread the door away from the jamb sufficiently to insert the pry bar.

Modern doors and jambs sometimes fit so well that it is impossible to even insert a wedge. In this case, a stainless-steel shim can be used. Force the shim into the narrow crevice between the door and its frame and attempt to push the bolt back.

Another way of opening a door is to use a so called Z-wire (fig. 44). This is a tool made from a stiff, thick wire, 25 to 30 centimeters long. Insert the Z-wire between the door and the jamb until the short end is all the way in. Then rotate it toward you at the top. This will cause the

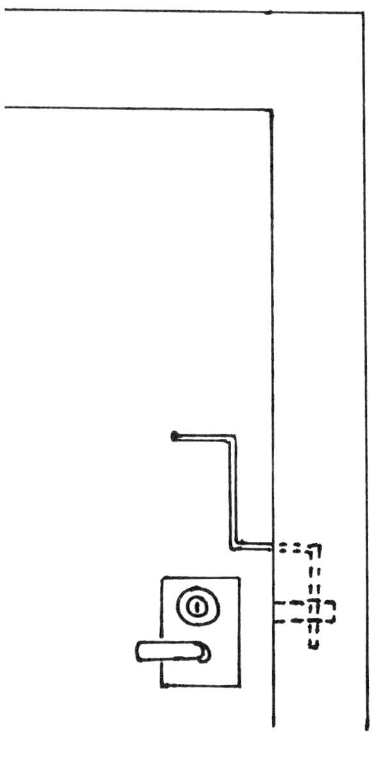

Figure 44. Opening a door with the Z-wire.

opposite end to rotate between the door and the jamb, contacting and hopefully retracting the bolt. When the bolt binds, exert pressure on the knob to force the door open.

Another problem sometimes encountered is a locked door chain. When you have managed to open the door, you will find that the door chain prevents the door from opening fully and that you cannot reach the chain in order to remove it. Fortunately, you can solve this problem by removing the chain with a rubber band. Reach inside and stick a tack in the door behind the chain assembly. Attach one end of the rubber band to the tack and the other end to the end of the chain (fig. 45). Make certain that the rubber band is taut. Then close the door, taking care not to lock it again. If this is difficult, secure the lock with adhesive tape so that the mechanism cannot work before you close the door.

When the door is closed, the rubber band will pull the chain back. If it is not completely pulled off the slide, shaking the door a little should do it.

The rubber band method is effective and easy to use, but it leaves an undesirable mark that can be seen easily on any well-kept door. It is then much better to use a bent coat

Figure 45. Unlocking a door chain with a rubber band and a tack.

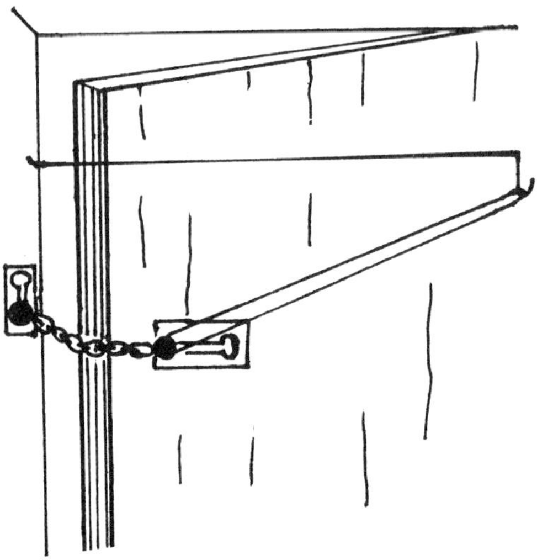

Figure 46. Unlocking a door chain with a rubber band and a bent coat hanger.

hanger to stretch the rubber band (fig. 46), rather than a tack. The coat hanger must be properly bent and long enough so that the door can be closed as far as possible.

An even easier method that can be used on doors with large enough space between the door and the jamb is to insert a thin wire to move the chain back.

If a transom has been left open or unlocked, there are a number of ways to enter. If the transom is completely open, simply crawl through, stepping on the door knob if necessary. If the transom is only partially open, however, it might be impossible to crawl through. Then lower a length of cord through the transom to form a loop which, when wrapped and drawn taut around the inside doorknob, might twist it enough to open the lock if you draw upwards on one of the two ends (fig. 47).

It is often easier to use two long pieces of string connected by a strip of rubber inner tubing or an electric cable covered by a strip of rubber than to use ordinary cord. The tubing should be 20 to 25 centimeters long. It is possible to open both a regular door knob and an auxiliary latch unit with this method, as long as a key is not required from the inside as well.

Yet another means of covert entry is to gently pry loose the molding around one of the door panels with a thin, flat chisel. If the panel can be removed without damaging it, the operative can crawl through the opening or at least reach through it and release the lock in the door.

It is also possible in some cases to open a door simply by bending it open by spreading the door frame. This can be done unless a very long bolt is used or the frame is very rigid. Take care that the door is not damaged, however, as this happens easily.

In certain countries, especially tropical ones, doorsteps are not very common. It might then happen that the owner of the house locks his front door from the inside and then leaves the key in the lock. An old but still reliable trick is to insert a paper sheet, an old newspaper, for

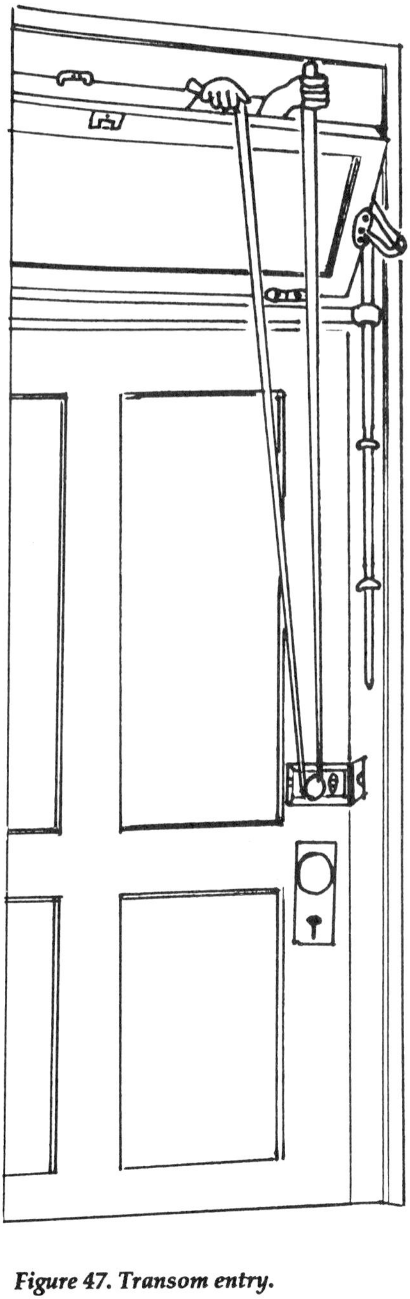

Figure 47. Transom entry.

instance, under the door. Then push the key out of the keyway with a wire. As the key falls down and lands on the piece of paper, you can easily pull both paper and key back and then use the key to open the door.

File Cabinets and Desk Drawers

There are various ways to open file cabinets and desk drawers, so you have to determine the best method based on the construction of the cabinet or drawer and the type of lock used in the object. With the exception of most sliding cabinet doors, however, the design is more or less standardized.

File cabinets are found in most offices. The locking device of such a cabinet relies upon a bolt that extends from the top of the lock body. The bolt manipulates the control bars that lock each drawer of the cabinet in place. As the bolt is spring-driven, it will extend out of the lock at all times, unless the lock plug is turned. The plug has a cam attached to the back, which rotates in conjunction with the plug. This cam works in a notch on the side of the bolt, thus operating the bolt (fig. 48). The cam can also be the actual lock bolt, especially in many drawer locks.

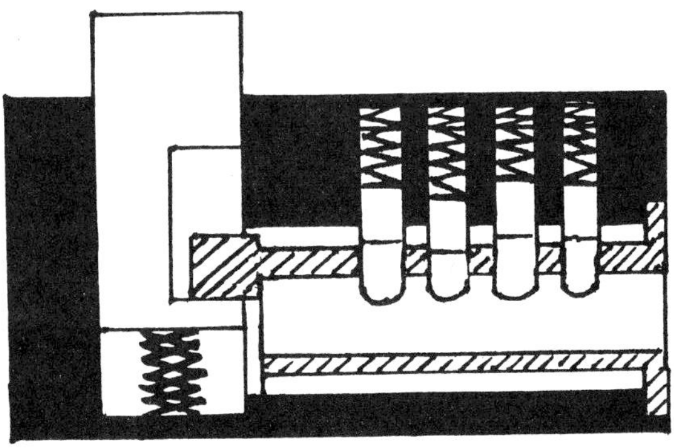

Figure 48. The cam of the plug operates the bolt in most file cabinets.

Locks of this type are generally not very secure. They usually have open keyways, with the keyway running from the face all the way through the body of the plug. This means that the lock can be opened most easily by jimmying. A jimmy is a pointed tool made of a thin strip of spring steel 3 to 5 millimeters wide. Slip the jimmy through the keyway and manipulate the bolt directly, pulling the bolt downward, without bothering to actually pick the lock (fig. 49).

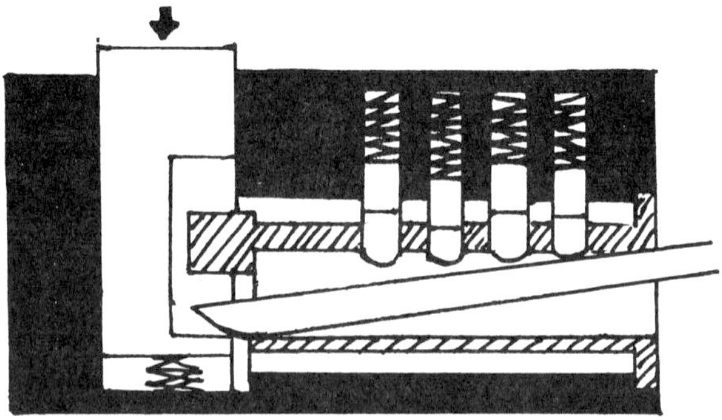

Figure 49. Pulling down the bolt with a jimmy.

Some contemporary locks of this type do contain some kind of protection against jimmying the lock open. For instance, the keyway might be blocked by a piece of metal or a pin so that no access to the bolt is allowed. These locks must be picked, or, alternatively, a rod of stiff wire not more than 5 or 6 millimeters in diameter and at least 30 centimeters long, with one end turned 90 degrees, can be inserted between the drawer and the cabinet face. The bolt mechanism can then be forced down with this wire (fig. 50). The rod should be turned counterclockwise in most cases.

Yet another method also involves prying the drawer open with, for instance, a thin piece of string steel or a wedge, far enough to allow direct manipulation of the bolt. A strip of steel, about 45 centimeters long and between 1 and 2 centimeters wide and 1/2 millimeter thick, or even an ordinary letter opener, can then be used between the drawer catch and the bolt mechanism to pull the drawer open. The opening tool will create a bridge for the drawer catch of the mechanism to ride upon and pass the bolt (fig.

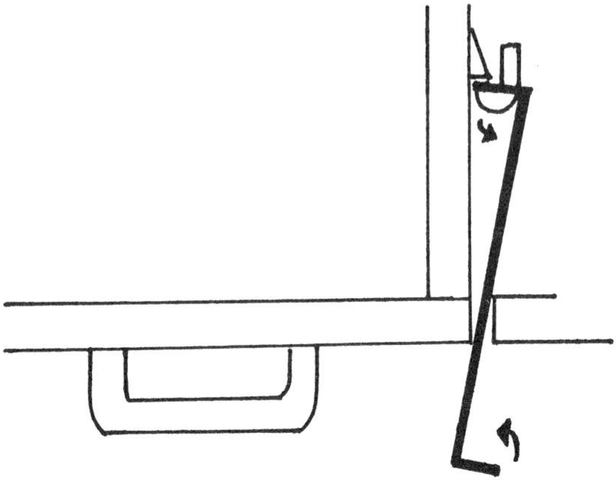

Figure 50. Forcing the bolt mechanism down to release the drawer catch.

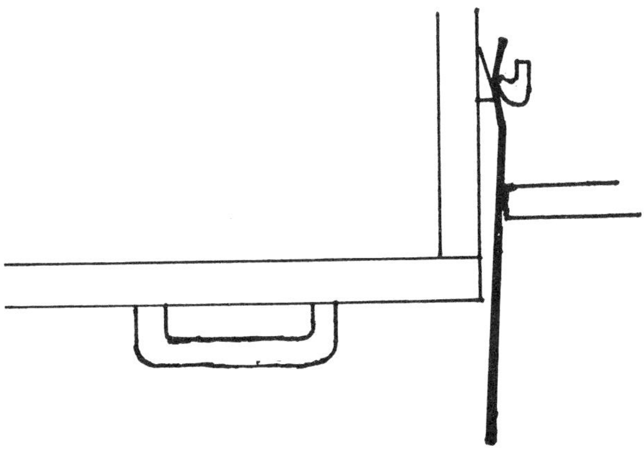

Figure 51. The drawer catch passes the bolt mechanism, riding on an inserted strip of steel.

51). Usually you must pull quite hard. Older types of file cabinets can be opened by simply spreading the file cabinet from its frame with a screwdriver, while a thin but strong tool, say a hacksaw blade about 30 centimeters long, is inserted in order to lift the bolt mechanism away from the drawer catch.

Certain types of file cabinets function in the same way, but with the mechanism effectively out of reach, hidden in the rear portion of the cabinet. Then it is sometimes possible to tip the file and in this way gain access to the mechanism, forcing it upward to release the drawers. If the cabinet is designed in this way, the mechanism can be seen protruding through the bottom partition of the cabinet.

Yet another method, one that works with certain types of file cabinets, is to tilt the cabinet backward about 15 to 20 centimeters off the floor and then suddenly release it so that one drawer catch will disengage itself from the catch on the mechanism rod. The drawer will then be released, and the operative will be able to open the remaining drawers by hand. Simply reach inside the cabinet and release the rod that keeps them in the locked position.

Certain file cabinets instead have gravity-type vertical engaging bolts. These can be released most effectively by inverting the entire structure.

Although most file cabinets have the lock in the same position, there are various ways of arranging the locking bar and locks in a group of desk drawers. This makes it important to choose the best way when it comes to opening a drawer. If all other methods fail, it is always possible to pick the lock itself. The best option, of course, is to find an identical cabinet and practice on it before the actual operation takes place.

Desks with locking drawers controlled from the center drawer can be opened in another way. Many of these are constructed so that there is a space between the back panel of the desk and the back of the desk drawers. The locking bar is usually designed to engage the desk by either upward or downward pressure. Whether it is upward or downward will depend on the style of bolt used in the design. The spring-loaded bolt will be pushed automatically into the locked position by the motion made when the locking drawer is closed. To open the side drawers, the operative must push the bolt up by hand from under the desk, reaching up between the back panel and the back of the desk drawers. When the bolt is raised, the hook catches will be released and the side drawers can be opened. The center drawer will have its own lock, of course, which might have to be picked open.

Other types of desks might require the use of a little force coupled with pulling outward on the center drawer. This will push the bolting mechanism downward just enough to open the various drawers.

The center drawer lock can be opened in the following way. First insert a screwdriver or another prying tool between the drawer and the underside of the desk. It might be necessary to protect the underside of the desk from scratching with some cardboard or thick tape. Then pry the drawer away from the desk top. Now use another, similar tool to pull the drawer outward to open it. Be careful, though, as this process might damage the drawer.

A final way of opening the lock, which is not really recommended as it will leave clear traces of the attempt, is to drill a small hole in the drawer above the lock. Insert a stiff piece of wire, such as a paper clip, into the hole to push down the plug retainer ring. This will pull the plug free of the lock, which will cause the bolt to drop down into the open position. The resulting hole can be partially hidden by inserting a wooden plug of the right material, but it will always be revealed by a careful investigation.

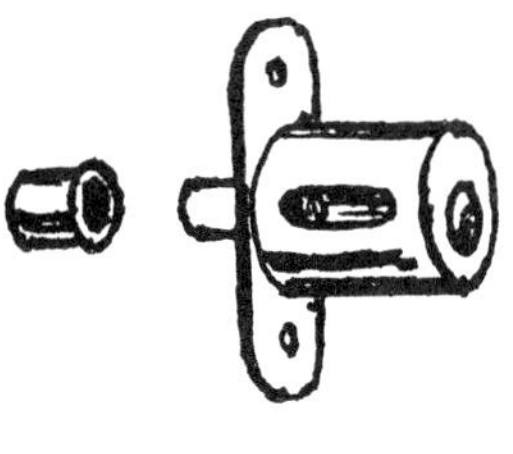

Figure 52. Sliding cabinet door plunger lock.

A sliding cabinet door is often locked by a plunger lock (fig. 52). Such a lock is mounted on the outside door. The bolt is a projection from the rear of the plug unit that engages a hole in the other sliding door of the cabinet. Neither door can then move, as they are locked against each other. The plug will be returned to the open position when the correct key is inserted into the lock and turned. The actual lock will be either a disc tumbler lock (in old buildings) or, more commonly, a pin tumbler lock. The easiest way to open these locks is to spread the two doors far enough apart to disengage the bolt. Alternatively, the lock can be picked in the ordinary way.

There are numerous ways of entering the average car. The door windows, the front and rear ventilation windows, the doors, and the trunk are all possible entrances. Sometimes the locks will be easy to pick, while at other times it will be easier to use some other method of entry. In this chapter we will look at some of these other methods. The methods used for picking the locks are described in Chapter 2.

The most vulnerable parts of the car are usually the windows. One reason for this is that most windows allow the door release push-button lever to be reached easily. This is especially true for the contemporary car models that have a single pane of glass. You can most easily reach and lift the push button with the help of a bent coat hanger. The coat hanger should be straightened out and then bent in a loop or a triangle on the end (fig. 53).

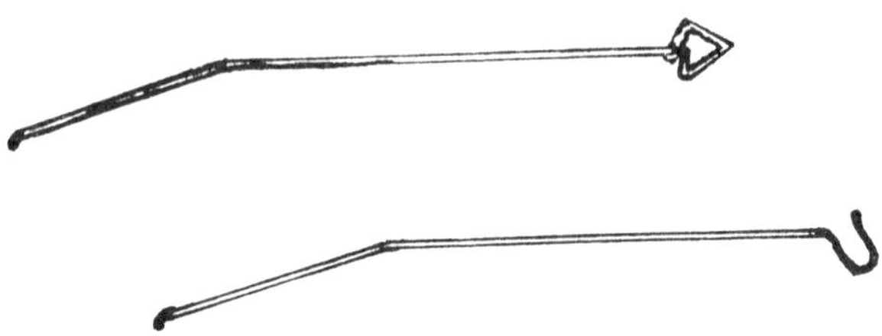

Figure 53. Coat hangers bent in loop and triangle.

First of all, as the window will usually be rolled up tight, force a paint scraper or a similar object between the edge of the window and the weather stripping (fig. 54). This will create an opening large enough to allow the coat hanger to be inserted in the resulting crevice. The loop or triangle can then be used to catch the push-button lever and pull it up. This will open the door.

Another possibility is to use a gun cleaning set instead of a coat hanger. As the gun cleaning set comes in sections, it is easy to make a rod long enough to reach across the inside of the car. In this case, always work on the window or door

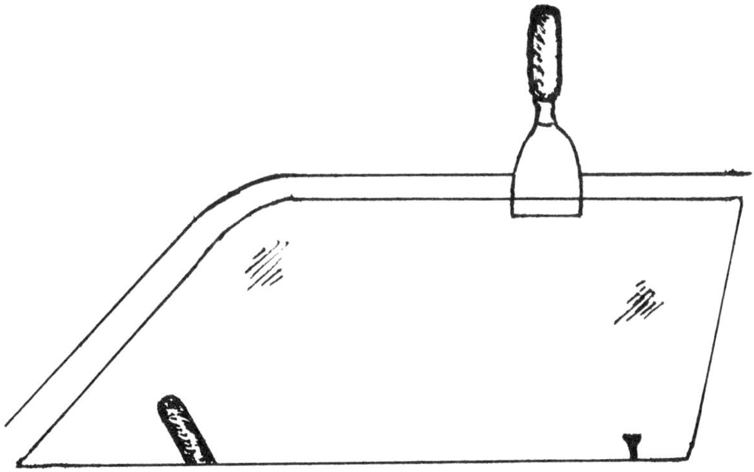

Figure 54. Paint scraper forced between the edge of the window and weather stripping.

opposite of the side where you insert the rod.

First of all, make a loop with some nylon line and fix it to the end of the rod on the slotted cleaning attachment included in the gun set (fig. 55). Then insert the rod and catch the lever or the door handle inside the car with the loop. Pull on the line and lift up the rod to raise the lever or the handle caught in the loop to the open position.

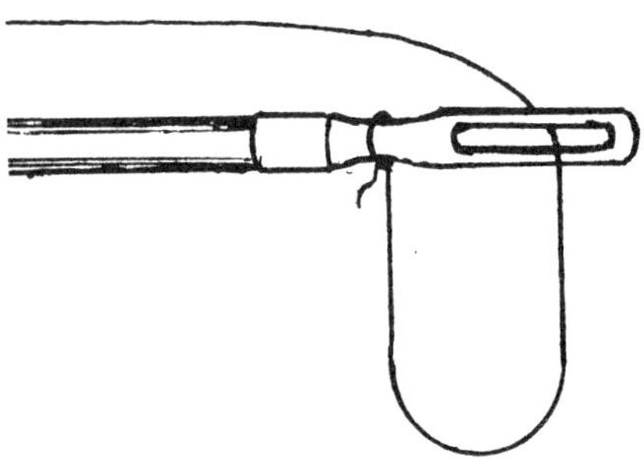

Figure 55. A loop of nylon line fixed to the slotted cleaning attachment.

If the car has a rear ventilation or wing window, this can also be used to gain entrance. These windows come in two types—those with a locking button on the swivel and those without. If there is no locking button, it is easy to force the swivel lock up and into the unlocked position. Simply insert a paint scraper or a similar object between the window and the frame. Then bend the tool slightly in order to make an opening wide enough to allow a thin piece of wire through. Take care to loop this wire around the swivel level and then pull it upward. This is accomplished most easily if the wire is slightly bent on the end so that it does not slip off the lever.

The problem, however, is that most cars nowadays have locking push buttons in addition to the swivel levers. There are special tools available to take care of them. Two tools are required, and they are inserted on different sides of the lock. Depress the button by pulling the first tool toward you, while twisting the second one slightly to push the lever into the unlocked position (fig. 56).

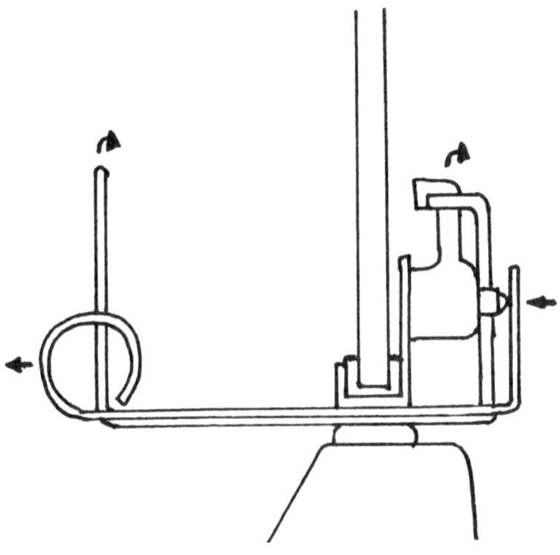

Figure 56. How to apply tools for opening rear ventilation windows with locking buttons.

The front ventilation window is another popular way of gaining entry to the vehicle. The reason is that it allows access to both the door handle and the window roller handle. Here, too, special tools are available that are very simple to use. Only pry the window slightly open, insert the tool, and turn the handle (fig. 57). Which type of tool is chosen generally depends on the amount of working space available, as their function is the same.

The trunk of the car is a completely different matter. As the same key is often used both for the door locks, the ignition, the trunk, and even the glove compartment if it is locked, it is often helpful to deal with all of these locks at

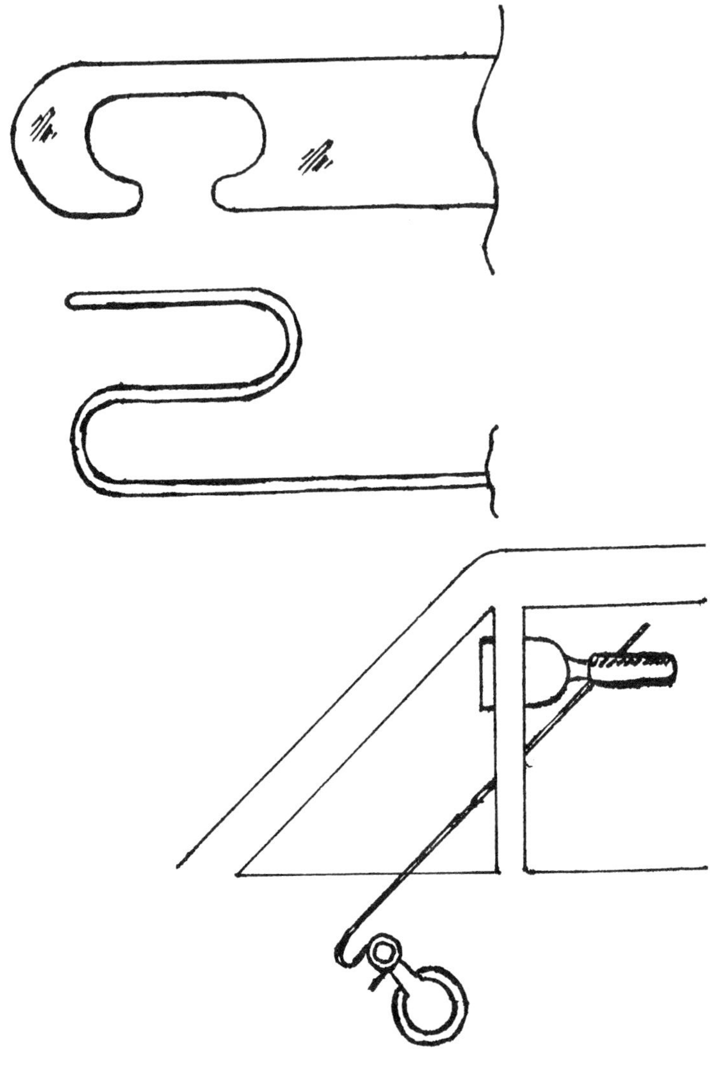

Figure 57. Tools for opening the front ventilation window.

the same time. Picking is the most common method of opening these locks, but it is often time-consuming. If you choose to drill the lock open, remember that many manufacturers have protected their locks by installing various steel plates or pins in front of the catch in order to make drilling more difficult. If drilling is required, it is often easier to break the trunk open or to drill beside the catch lock and then manipulate the catch with a bent piece of wire. In either case, the intrusion will naturally leave very clear marks on the car.

A few trunk or tailgate cylinder locks, as well as many glove compartment locks, can be opened in a much easier way. Examine the lock to see if this method is possible. A simple lock that is possible to open in this way is designed to be secured with a retainer accessible from the front of the lock. This lock can be opened by inserting a special L-shaped tool, in effect a 5- to 6-millimeter hook in the end of a piece of stiff wire, through the keyway (fig. 58). The retainer can then be pulled down and worked free. Then the entire plug can be pulled out of the lock and removed. To do this, force the retainer (usually installed with its open ends toward the passenger door) toward the center of the car to disengage it. When the plug has been extracted, the catch mechanism can be pushed back and released with the help of any pointed tool.

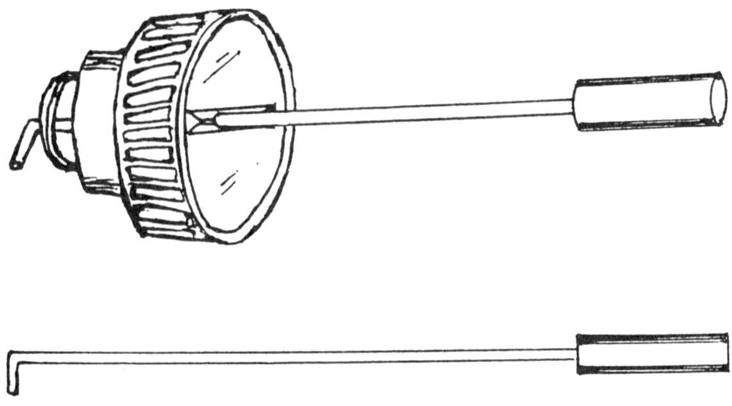

Figure 58. Removing the cylinder from the outside with the L-shaped tool.

There are, in fact, two different locks that can be opened with such an L-shaped wire. The lock in which the plug can be removed has already been described. The other lock is of an even simpler construction. By inserting the wire into the keyway, you can engage the catch mechanism itself and open it by merely pushing it downward.

CHAPTER 5

Methods of Forced Entry

During a forced entry, the actual lock is usually not the target of the intruder. The other components of the lock set and the door (i.e., the striking plate, the hinges, the door frame, and the panels in the door) are generally much weaker. Even the wall next to the door might be weaker and thus more vulnerable than the door itself. The same might apply for the roof or ceiling, or even the floor.

In most buildings built in a warm climate, the walls lack insulating materials. The wall very often consists only of an empty shell, extremely easy to break through with heavy-duty tools. Floors and ceilings are notoriously weak in most countries, whether the climate is warm or cold.

All of these facts help to facilitate a forced entry. The level of force employed in the operation can vary considerably, depending on the circumstances. Sometimes, a low degree of force turns out to be enough. In a few cases, however, the door or the wall area around the main door are smashed through by ramming the building with a car or a heavy truck. A tractor might be used for the same purpose. These are very efficient and quick, if somewhat noisy, approaches to forced entry. Less violence might be sufficient, but in every forced entry operation, it is important to choose the most expeditious method.

The door is often the weak link in the protection of the building. The simple fact that the door has to be opened frequently ensures that it cannot be made completely safe. It is important to determine which is the weakest part of the door. If the door is massive, then the lock will be the weakest part, but in fact it is much more common that the door itself is of a weaker construction than the lock installed in it.

If a lock is installed in the wrong way, however, or simply installed carelessly, it may be easier to break through the locking mechanism. Watch for major gaps between the door and its frame. If one of these gaps is large enough, it might be possible to simply insert some suitable object or prying tool through it and wrench the door open.

A cylinder lock is incorrectly installed if the cylinder protrudes more than 2 millimeters. If this is so, then the lock can be forced open by either pulling out the cylinder with heavy-duty tongs or drilling through the now exposed weaker side of the cylinder in order to destroy the locking mechanism.

A lock can almost always be drilled open. In a very few cases, the lock can then be replaced without the owner noticing, at least if it is a cylinder lock. If only the cylinder plug is drilled, the cylinder itself can be saved. Only the inner core needs then to be replaced. However, it is not possible to reconstruct the key afterward. The core has to be a new one, although the external appearance will be the same. When a cylinder lock is drilled open, the lower sets of pins just below the shear line are destroyed, which allows the plug to be turned, as the upper pins can be kept above the shear line with a wire inserted into the lock. There are also various other ways of drilling open a lock, but these will all destroy parts of the lock set that are exposed and clearly visible. The intrusion is therefore obvious and extremely conspicuous. If force of this kind has to be used, then it is better to plan a regular forced entry and simply wreck the entire door.

The area *around* the lock is also a good choice for breaking open. If, for instance, two mortise locks are installed in the same door, the locks should be at least 40 centimeters apart or the door structure will be weakened significantly. Even though the locks might be strong, the door is then easy to break.

In other cases, the screws attaching the lock to the door might be visible, and they might even be possible to remove from the outside. This is another major construction error that makes it easy to force the door open.

Some doors have been reinforced with steel plates fitted on the most exposed parts. If these plates are thick enough and the frame is not significantly weaker, they will cause problems during a forced entry. A carborundum wheel might be necessary to cut through the steel.

Doors with rebated wooden panels and most other types of panelled doors are very weak and can easily be broken with a hefty kick or a hacksaw blade. Aim for the panel next to the lock so that you can reach in with your hand and unlock the door. Be aware, however, that some of these doors have been reinforced by steel sheeting on the inside rather than on the outside. If this is the case, the door might have to be treated as a steel door, despite its inviting exterior appearance.

Some front doors do, in fact, have very flimsy lower panels. They are not only easy to break in through, but they can also be used to push larger objects out of the building. This is a technique often employed by burglars. However, when you are attacking one of these doors, make certain that the panels have not been reinforced on the inside by stronger and thicker wood panels. Such reinforcements can sometimes be detected by the appearance of a number of screws. Wood reinforcements, unlike steel reinforcements, are possible to break through with ordinary tools, but the job will take time and create noise.

Patio doors and sliding doors, especially those with either wooden or aluminum frames, are usually very easy to break through by simply lifting the doors out of the fitting. A patio door lock is a security device designed to bolt the sliding door to its frame. It is never very strong, though. Some patio doors are even made of plastic today.

Glazed doors are extremely vulnerable. It is not necessary to break a big hole, however. A small hole, sufficient to allow the operative to reach in and open the lock, is quite sufficient. A small area of glass is broken more easily—and safely—than larger ones. A French door, for instance, is likely to open outward, and the latch can be reached and opened easily by breaking one of the small panes of glass in the door. Such a small pane can be broken quietly, without the risk of the neighbors hearing it.

Most doors with a mail slot are very vulnerable to an attack through this opening. Simply put a crowbar through the mail slot and break open the part of the door below the mail slot. This job takes only a few seconds. The latch can sometimes also be reached through the mail slot itself. Special equipment can be made for this purpose.

No door is stronger than its frame. A softwood frame is easily broken. Exposed hinge pins can also be worked upon to open the door. If a wedge was not fitted between the frame and the wall opposite the lock when the house was constructed, then the frame will be weakened significantly. In this case, the frame can often be simply spread away from the door, thus allowing the door to be opened without bothering with the lock mechanism. This is even true of most steel frames, at least those not thicker than 2 millimeters.

Most outward-opening doors can be opened by knocking out the hinge pins and then prying the door open on the hinge side. Alternatively, if the pins cannot be knocked out, the hinges can be sawed off. Therefore, a strong door

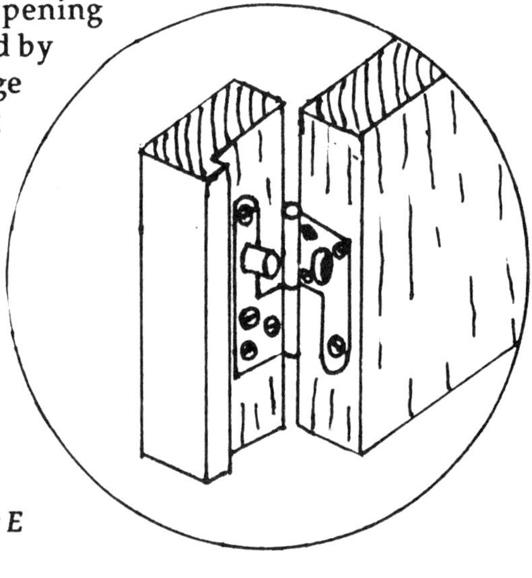

Figure 59. Hinge bolt.

may also be protected by a hinge bolt (fig. 59) fitted in the hinge edge of the door close to the hinge positions. These are studs set into the hinge edge of the door which engage into corresponding sockets, or recesses, in the frame whenever the door is closed. The door cannot then be forced open or lifted simply by sawing off the exposed hinges. Such a door can only be forced by removing either the lock or the frame area around the hinges and the hinge bolts. Hinge bolts are always installed in pairs, one usually below the top hinge and the other above the bottom hinge. Mortise rack bolts are sometimes used for a similar purpose, as for external doors. Such bolts are fitted at the top and bottom of a door and can be locked from the inside with a universal splined key (fig. 60). The lockable bolt is the type most commonly used today. Once again, these bolts can only be locked from the inside. The mortise bolt is a dead bolt, so it cannot be released without a key.

It is worth mentioning that in older houses some doors and French windows are usually secured with an espagnolette bolt that extends the full length of the door and consists of two vertical sliding bolts, one covering the top half of the door or window and the other covering the bottom half. Both are operated by a central handle, which is often lockable. This works according to roughly the same system and is used to keep the door in place even if the hinges are destroyed.

Figure 60. Mortise rack bolt, locked from the inside.

Barrel bolts are used in many buildings for a similar purpose. These bolts are screwed onto the surface of the door and usually shoot into

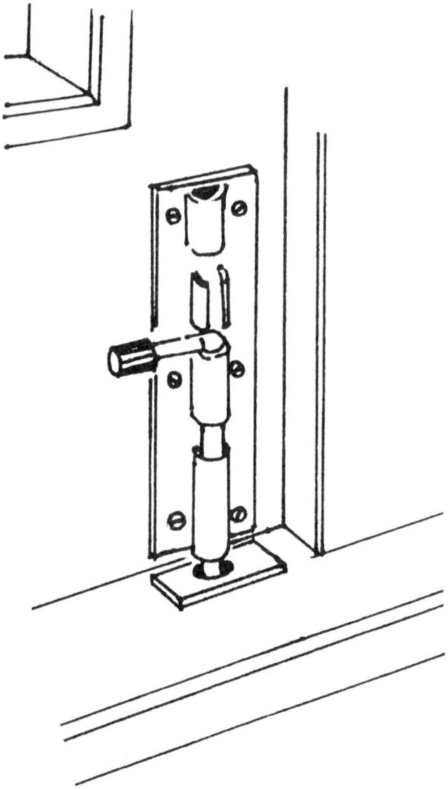

Figure 61. Barrel bolt.

a staple on the door frame. They are usually not fitted with locks. If they are used with a staple, the staple is the weak point, and it can often be forced. If the bolt instead is mounted so that it shoots into the head or sill of the frame, it is much stronger (fig. 61). Sometimes two barrel bolts are used, one for the head and one for the sill. Door chains and door limiters (fig. 62) are devices that pose similar problems for the intruder, but only if somebody is at home and has put the chain or limiter in position. The limiter is similar to the door chain, but it uses a sliding-rod device instead of a chain. Both these devices are supposed to keep the door safely and effectively closed, even if it has to be slightly opened, such as when receiving a small parcel or a letter. The chain will allow the door to open to about 5 centimeters, but no more.

Certain door chains do in fact have a key-operated lockable staple on the door frame, which allows the occupant to use the chain even when leaving the house. On return, he can open the door sufficiently to unlock the chain from the staple. These door chains can cause a problem for an intruder, unless the staple is weak.

As a matter of fact, many door chains are very weak.

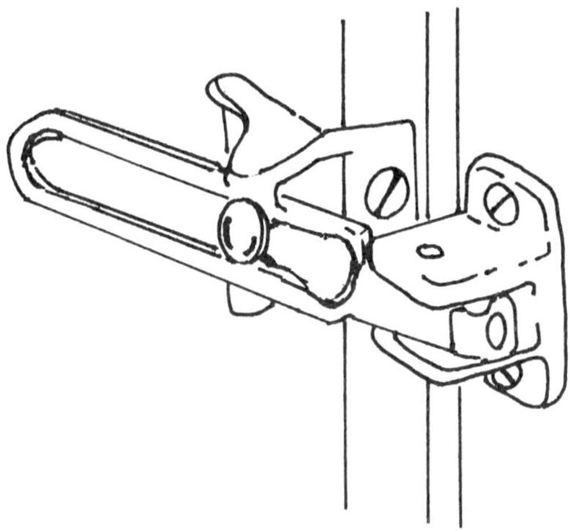

Figure 62. Door limiter.

There have been cases when door chains have been broken as a result of someone merely opening the door without first removing the chain.

Sometimes the door will not be locked properly because of a faulty door check. A door check, or door closer, consists of a heavy spring and arm coupled to an air or oil cylinder that automatically closes the door. The door check also controls the speed at which the door closes. If the door check is not working properly, there is a good chance that the door remains unlocked.

It is also important to plan the break-in properly. In some buildings, for instance, the front door might be very difficult to break through, but a garage will be attached to the side of the house. Inside the garage there will then be only a lightly protected internal door linking the garage with the home. This door might well turn out to be a much easier target than the front door.

Windows

Windows always present a special problem. There are numerous types of window locks for casement windows, sash windows, and most other types. These locks can generally be picked from the inside, as they are extremely simple and sometimes even all use the same universal splined key. From the outside, however, the only means of entry is to break the window or lock open. Fortunately, the very construction of a window makes it relatively weak and easy to break. A forced entry through the window is almost always easy. If the window is only closed and not locked, the easiest way to gain entry is to simply smash a small pane to allow one's hand through to release the catch. The window can then be opened.

Window locks will unfortunately prevent the window from being opened in this way. Even if the window is smashed, the frame will remain closed. Climbing through this frame will necessitate noisily breaking a large amount of glass; this is quite dangerous, since the operative is likely to cut himself. Naturally, both these problems should be avoided if at all possible.

In many houses in which the windows are locked, the owner will keep the key to the lock of one window close to the window in case there is a sudden fire. If this key can be recovered or reached, then the window is effectively unlocked. This should be remembered if a window entry is being considered.

Louver windows are notoriously easy to force, especially from a flat roof, such as on a garage, that offers easy access to the window. The louver window can simply be levered out of its frame, unless it is glued very firmly with epoxy resin. In this way, whole strips of glass can be removed from the metal clips. Modern louver windows might have locking devices installed (fig. 63). Sometimes the louver blades are also made of laminated safety glass. Still, a louver window

Figure 63. Louver window lock.

will be a good point at which to attempt a forced entry.

Metal window frames in older houses are generally made from steel or galvanized steel and, consequently, are quite difficult to break through. Modern aluminum-framed windows, however, are made of such a thin, soft aluminum alloy that the frames can be distorted easily with a lever of some kind. The self-tapping (self-threading) screws used in these constructions tend to pull out easily, and the locks, if any, are often of a very poor quality. This is especially true of horizontal sliding windows, which work by sliding on an aluminum track. It is often relatively easy to lift such a window out of its track, using the same technique described above for lifting a sliding door.

The important thing to consider when it comes to entering through a window is the type of glass used in the window. There are several different types of glass available. The lowest quality is sheet glass, a cheap glass with imperfections. It is also often referred to as horticultural glass, as it is used for glazing greenhouses. It is generally not used for domestic glazing. A sheet is 3 millimeters in thickness and will break easily.

The standard material for domestic glazing is float glass, which is flat and free from imperfections. It generally comes in thicknesses ranging from 3 to 10 millimeters. Glass that is 4 millimeters thick is very common, but the thickness is always related to the size of the window; a larger window requires thicker glass.

Wired glass is a rolled glass in which a wire mesh is embedded during manufacture. It is 6 millimeters thick and can come both with square or diamond-pattern mesh. If the wired glass is hit by something, the wire will hold the glass together. Its resistance to impact is high, so it is often used where fire-resistance is important, as well as on glazed roofs on which snow and ice are likely to fall. Wired glass is generally not used for security glazing, however, as the mesh can be broken through once the glass has been smashed.

Laminated glass is true safety glass. It cracks but does not break under impact. This glass is formed by two sheets of float glass with a thin sheet of crystal-clear plastic sandwiched between them. It is very strong. Although the glass will crack under heavy impact, the plastic will hold it together very firmly. This is the most common type of security glass.

Tempered glass, also known as toughened glass or armor-plate glass, is heat-toughened safety glass that is both impact- and fire-resistant. When it does break, it shatters into numerous but harmless pieces with no sharp edges. Tempered glass is four to five times stronger than ordinary glass of the same thickness.

Many other types of glass, such as patterned glass, solar control glass, etc., are also in common use but are not encountered very often during a break-in. They fulfill no special security purpose. Of these various types of glass, only wired, laminated, or tempered glass will resist a sledge hammer, and the wired glass can still be broken through with the help of other tools. Most types of glass can be broken eventually, but doing so is sometimes too time-consuming and noisy to be truly efficient.

The intruder breaking into a window must also look out for venetian blinds. These create two different problems. First of all, it is often very difficult to see through them to determine whether a room is occupied or not. Secondly, breaking through them makes a terrible noise. For these reasons, windows with venetian blinds should be avoided.

Security Bars

Another problem for the intruder will be the existence of security bars. Such bars, used as reinforcement next to an existing door or as a decorative grille for a window, are very difficult to remove, except through brute force or the noisy use of a carborundum wheel. Gril'es, especially large sliding grilles (fig. 64), are commonly used in many countries for protecting large areas of glass or doorways. Other grilles are of the detachable type, locked in place with only fixed locks or padlocks. If this is the case, the locks are the weak links in the construction.

Some older buildings rely on solid wooden shutters

Figure 64. Sliding grille.

instead of steel grilles. Wooden shutters of the louvered type are especially easy to break through, so they present no special problem.

Safe Cracking

There are many different types of safes. Some of them are built to be resistant to burglars, while others are resistant to fire. Some are designed to resist both. Most safes can

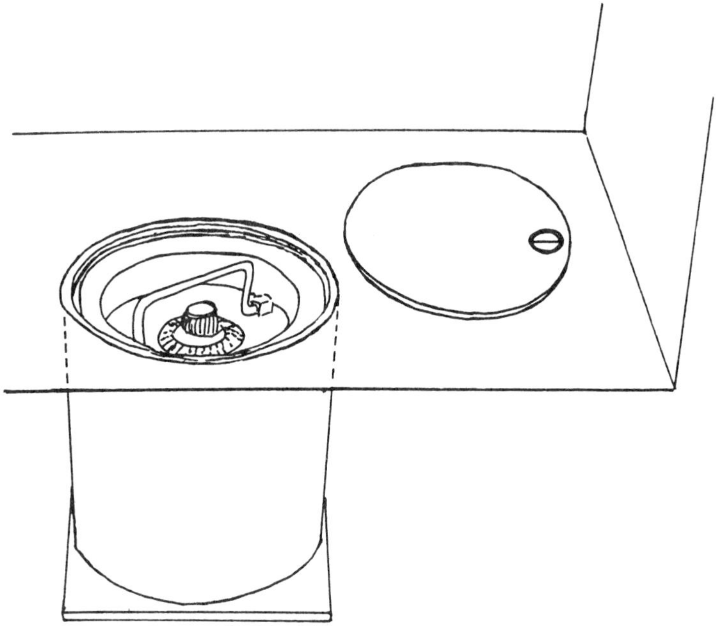

Figure 65. Floor safe.

be cracked without too much trouble, although the contemporary models are more difficult than the older ones. The main problem with cracking a safe is that it creates a large amount of noise. Furthermore, most safes are too heavy to move from their location, so the work must take place on the premises. If the safe is small and light enough and not fixed to the wall or to the floor, then by all means remove it and open it at leisure.

The most reliable safe in a house is the floor safe, as this type of safe is out of sight and can be permanently set in a reinforced concrete floor (fig. 65). A floor safe is recessed into the floor so that the small but strong lid is just below or at the floor level, and it is made of thick steel. Different sizes are available, depending on the depth available under the floor, but the opening will almost always be relatively small.

A safe of this kind is generally positioned in a corner

near the wall, so that it can be covered with the carpet or linoleum and still reached with a minimum of inconvenience. A corner location will also give the intruder less room to work, even if he happens to discover the safe.

A floor safe will not be found in a room where the floor is likely to get wet, such as in a bathroom or a laundry room. Even though it is usually hidden under the carpet, other hiding places should not be disregarded, such as a false floor in the base of a cupboard, for instance. Such a cupboard can be located easily, even in a room with a tiled floor.

A floor safe can also be found fitted in a suspended

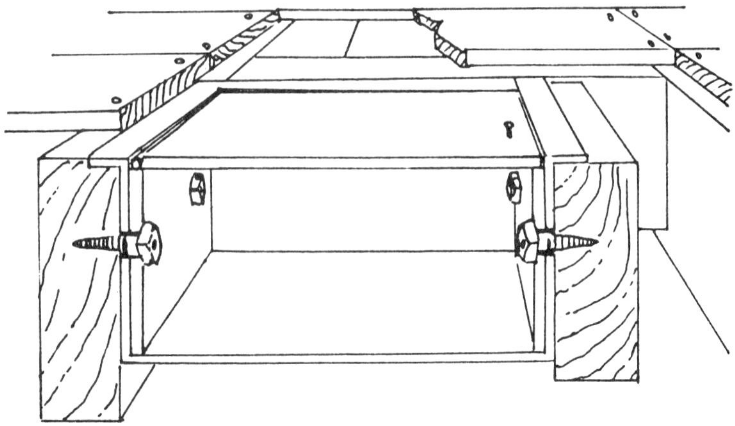

Figure 66. Floor safe in a wood floor.

wood floor (fig. 66). Such safes are in effect secure boxes, often with combination locks, which fit between adjacent joists and are bolted or screwed to them. The screws or bolts are fastened from the inside of the safe. The floorboards will be removable in order to allow access to the safe. The safe itself generally is covered with a sheet of plywood or hardboard to bring the surface level with the surrounding floorboards.

In an apartment complex, the equivalent of the floor safe is the wall safe (fig. 67), which is also easily concealed. A wall safe is a small security box that is set into the wall, replacing one or more existing bricks. The size of the wall safe is measured by the number of bricks it replaces. Wall safes are therefore commonly one, two, or three bricks high and one brick deep, although double-depth models are also available.

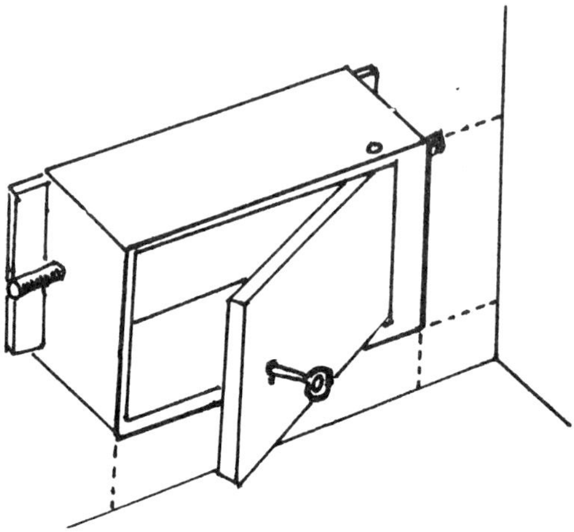

Figure 67. Wall safe.

Wall safes are sometimes very cleverly hidden, although it is amazing how often they will actually be positioned behind a painting or another picture. Therefore, this is the first place to look for a wall safe. If the wall safe is not in such a place, the owner has been clever, and it could be anywhere. There is even a commercial wall safe the size of a cash box that is disguised to look like an electrical power socket (fig. 68).

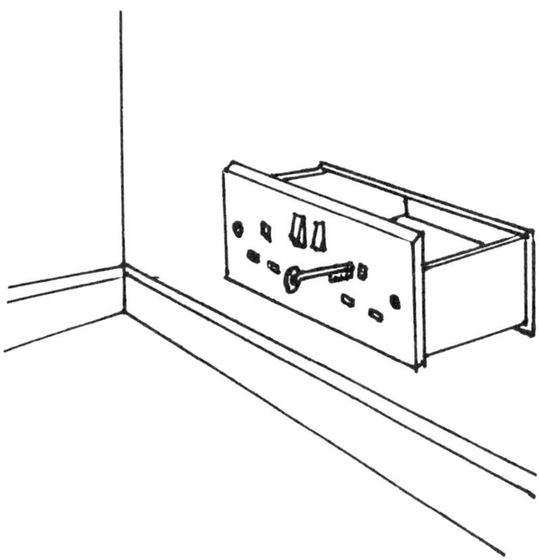

Figure 68. Small wall safe disguised as electrical power socket.

A wall safe is easier to breach than a floor safe, as the latter is more difficult to lever out of position. It is also quite easy to knock out a masonry wall, especially in an older house where the mortar is crumbly. The bricks around the safe can be chopped out with a bolster chisel. The wall safe is then removed and brought to a safe place to be opened.

Certain buildings might have a strong room instead of a safe. These are sometimes difficult to break open, but the principles are the same as with ordinary safes. A strong room is generally less secure than a bank vault, which is very difficult to force open.

The methods suitable for cracking ordinary safes should be helpful for breaking into a bank vault, too. Another interesting point about a bank vault is that it often has a hidden emergency entrance in the form of a hatch. This is to ensure access even if the main door has become impossible to open because of a failed burglary or a mechanical problem. This hatch will of course also be heavily protected, but as it is smaller and usually hidden behind a steel plate, it is not as protected as the main door.

Free-standing safes, especially if not prohibitively heavy, are often secured to the floor, and sometimes to the wall, too. If the safe is not secured in this way, the best option is to simply remove it and crack it open at leisure in some safe spot. All safes are cracked most easily with the help of a carborundum wheel. The disadvantages of this method are that a steady supply of electricity is required and it makes a terrible noise. Fortunately, sufficient power is in fact available in most houses, although it is prudent to bring an extra set of fuses in case the regular ones have been removed in order to prevent this method from being used on the safe.

Many types of safes rely upon combination locks. These can be manipulated to the open position, but it is difficult. The technique is described in Chapter 2. Combination locks can also be opened by drilling, however. The lock cannot be opened completely by simply drilling, but the drilled holes will help the intruder manipulate the wheels of the combination lock.

If this method is resorted to, drill two 3-millimeter holes

in the back of the lock. Then turn the dial and observe what is happening inside the lock through these holes. The gate can be seen through the hole with the help of a flashlight, or a piece of piano wire can be inserted through the hole as a probe. When you have found the gate and aligned it with the hole, you can note the number on the face of the combination dial. Then determine the distance, expressed in divisions on the dial, between the bolt and the gate aligned with the hole. Subtract this distance from your reading. This will give you the combination number for that particular wheel. After obtaining the first number of the combination, reverse the rotation of the dial and repeat the process with the second wheel. The combination of the third and the fourth wheels (if the latter is present) can be determined in the same way.

There are numerous cruder ways of cracking a safe. Explosives, for instance, can be used. Then it is most common to drill a hole above the dial and insert a finger of a glove or a small plastic bag with explosives. An explosion here will destroy the lock mechanism. However, it is by no means guaranteed that the door will open because of this, as it is fairly common for the door to jam during the explosion. This method is therefore not completely reliable.

Another method, almost as old, is to use nitroglycerine. First drive a steel wedge into the top seam of the door. This allows the nitroglycerine to seep around the inner edge of the door so that the door will be blown off the safe when the nitroglycerine is detonated.

None of the methods requiring explosives are really recommended, as they are both dangerous and create too much alarming noise. A neighbor might be disturbed by the sound of a carborundum wheel, but this alone will not be sufficient to make him or her call the police. After all, many repairmen use this tool. An explosion, however, is quite a different matter.

Another way of cracking older, inferior types of safes is to drill a hole in the corner of the front plate of the door. This plate can then be torn away with the help of a long crowbar in order to expose the lock mechanism. The lock mechanism can then be manipulated open easily. This method is completely ineffective against modern safes, however, as their doors are massive.

In fact, most of the older methods are obsolete, as the safe doors are laminated with hard steel and beryllium-copper plates. Drilling, for instance, becomes almost impossible. One method that still works is generally known as a "torch job," in which an oxyacetylene torch is used to cut through the safe. This requires bulky equipment and special training. It is not really common nowadays, as a carborundum wheel will do the same job in an easier and safer way.

Most older types of safes relied on fairly thin metal walls, padded by an insulating material. However, after several years have passed, the insulating material will have compressed, eventually leaving the upper parts of the walls completely empty and therefore very easy to break through.

Another curiosity should be mentioned here. Some companies now market what they call a bionic safe. This is a safe with an integral alarm system, as well as a few other protection devices. The sensor is usually an inertia sensor that will activate a siren and at the same time, through an automatic dialer connected to the nearest telephone, dial a programmed telephone number. If the safe is broken despite these precautions, a self-contained explosion will destroy all materials inside. (It is designed to cause no harm to any people or property in the vicinity, however.) Furthermore, if the safe is opened without authorization, a sudden burst of high-powered light will be emitted. This is to disable the intruder by stunning and blinding him temporarily. This type of safe is not yet in widespread use.

As a final note, it should be remembered that many safes are designed with an outer construction that imitates high-quality wood paneling. This is both to hide the safe and to make it fit inconspicuously into a home or office. The best protection has always been to hide the safe well. Therefore, when searching a home or an office, be careful to check all possible hiding places, however small, for a safe.

CHAPTER 6

Alarm Systems, Sensors, and How to Avoid Them

According to data from the British police, as many as 98.6 percent of all alarm calls are false. Similar data can be found in most other countries. Clearly, a single-alarm call may not be such a great threat to an entry operation. However, alarm calls must be avoided at all costs—especially multiple calls from the same system.

The basic intrusion alarm system consists of:

- a control unit, generally installed within easy reach of the main means of exit and entry, usually the front door. This is the brain of the alarm system
- one or more warning devices, such as sounder boxes containing warning bells or sirens, and/or strobe lights, often fixed to the wall on the outside of the house
- one or more detection devices, or sensors

Other major components include devices for arming and disarming the alarm system, automated dialing equipment, power supply such as batteries, and wiring between the various components.

The control unit is housed in a protective box, generally made of metal, and is situated in a central location in the area to be protected. An indoor closet might be used for this purpose.

Warning devices of these types are sometimes referred to as alarms or annunciators, although the word alarm is most often used for audible warning devices. Additional devices may be fixed inside the house, mainly as a psychological disturbance to the intruder, and so-called silent alarms are also a possibility. In the latter case, a remote signaling system will be used.

The sensor is the device that relays information to the control unit. If the control unit is the equivalent of the human brain, then the sensors are similar to the senses of the human body. A sensor, or detector, is a scanning and screening device. Its effective range is called the detection zone. The sensors are of different kinds and are commonly divided into three lines of defense. The sensors are connected to the control unit, which, after receiving a warning from a sensor, transmits the alarm to the warning device, which will sound the alarm. The various types of sensors will be detailed in Chapter 7.

There are three lines of defense, but in fact four different kinds of alarm protection:

- external alarms
- perimeter alarms
- trap alarms
- deliberately operated alarms

External alarms aim to detect an intruder as early as possible before he actually reaches the main building. It relies on sensors located in the grounds or on the boundary wall or fence. As these sensors are often susceptible to false alarms, they will generally be monitored by a private security staff and are therefore only to be expected in extremely rich neighborhoods or corporate or government installations.

Photoelectric cells are frequently used for this purpose. They can be positioned to protect the entire perimeter. Other common sensors are the microwave fence and the field effect detectors. Special barrier or fence detectors are also used in certain high-risk installations. Another device used for external alarms is the geophone.

Perimeter alarms are designed to protect the shell of the building (i.e., the walls, doors, and windows). The perimeter sensors will detect the intruder as soon as he breaks into the building. Perimeter alarms are very common and are often used in conjunction with trap alarms.

There are numerous types of perimeter alarm sensors. Magnetic reed contacts, photoelectric cells, glass breakage detectors, video detectors, vibration detectors, inertia sensors, infrasound sensors, field effect sensors, plunger switches, and pressure mats are commonly used sensors in the perimeter alarm. Window foil or prefabricated window foil and wire contacts are also used in some older systems.

Trap alarms are detection devices installed at strategic locations within the house to detect an intruder after he has already entered the building. They are also commonly used to protect individual items of great value or importance.

Detectors used as trap alarms commonly include passive infrared detectors, microwave movement detectors, ultrasonic movement detectors, photoelectric cells, magnetic reed switches, pressure mats, light detectors, and video detectors. Other sensors, such as field effect sensors, sound detectors, and heat detectors (not to be confused with the infrared detectors) also fall within this category. Ionization detectors, when they come into general use, will also belong to this group.

Trap alarms also include special alarms used for guarding specific objects, such as valuable paintings, computers, and so on. These alarms can be of many types, but it is common to run normally closed circuits (this will be explained below) incorporated into the main cable to the electronic device, or affix them somehow to the object of concern. When the wiring is cut or pulled from the wall socket, whether or not the power is on, the alarm will sound.

Finally, deliberately operated alarms are also known as panic buttons. They are found in banks, for instance, but also in many private homes. Such buttons can frequently be found in the bedroom and just inside the front door. Deliberately operated alarm systems will be described in Chapter 8.

Basically, there are two main types of alarm system installations. Most systems consist of separate parts, but nowadays an increasing number of systems are self-contained. A self-contained alarm system contains the control unit, the sensor, and the warning device in the same, easily installed unit. Connections on the back of the unit allow the attachment of separate sensors and external warning devices. The self-contained systems have advantages and disadvantages. This is also true of the separate-components systems, however, so it is prudent to plan well ahead when installing an alarm system.

The separate-components systems are highly adaptable and expandable and can therefore be used in any size or type of building. Additional components can also be installed at a later time. The self-contained systems are more easily installed, however. Generally, they are of the tabletop variety and need only be positioned in a room and plugged in to be ready to use. Self-contained systems are also easy to move, both from different rooms in the same building and from one building to another. For these reasons, self-contained systems are more commonly used by people who rent their homes or offices, while homeowners and large companies generally use separate-components systems.

However, there is another, more significant difference between these two types of systems. The sensors used with most separate-components systems are usually designed as perimeter defense (i.e., to detect intruders and sound the alarm before they manage to enter the premises). Older systems thus require an externally mounted key switch, which serves as a remote arming and disarming device. The self-contained alarm systems, on the other hand, are generally designed to detect an intruder after he has already entered the building. Instead of an externally mounted switch, the system incorporates a time delay, which allows a certain time, often fifteen to thirty seconds, to enter the building and turn off the alarm. The alarm will be sounded only if the system is not turned off in time. It must be remembered, however, that self-contained alarm systems can also be connected to perimeter alarm components. If this option is used, the self-contained alarm functions as a hybrid control unit, relying on both its own sensor and one or more external ones.

In fact, many self-contained alarm systems can also be used in conjunction with separate-components alarm systems without linking them into the same circuits. The self-contained systems are then used in those areas where it is difficult or impossible to place sensors linked by wire to the control unit in the main system.

Otherwise, self-contained systems are used most commonly in apartments, especially those in which the owner prohibits the tenants from installing permanent alarm systems.

Self-contained alarm systems include:

- passive infrared motion detector units
- ultrasonic motion detector systems
- microwave motion detector systems
- self-contained window or door alarms
- infrasound detectors

Self-contained fire alarms are very common in many different types of buildings. All other types of sensors can generally be found only in separate-components alarm systems.

Most large alarm systems consisting of more than one sensor are designed to be split up into zones, whereby different areas of the building are controlled by different circuits. The main advantage of this is that the occupants of the building can choose to activate all, some, or none of the different parts of the system at any given time. For instance, a bedroom can be excluded so that the occupant can move around there without setting the alarm already activated in the hall near the front door.

Each circuit is made up of a series of contacts, all located within the same zone. The alarm will be triggered when any of these contacts is broken. It is common for several different types of sensors to be installed on each of these circuits.

However, all types of sensors, when they are designed to be connected to a control unit, operate as simple switch-

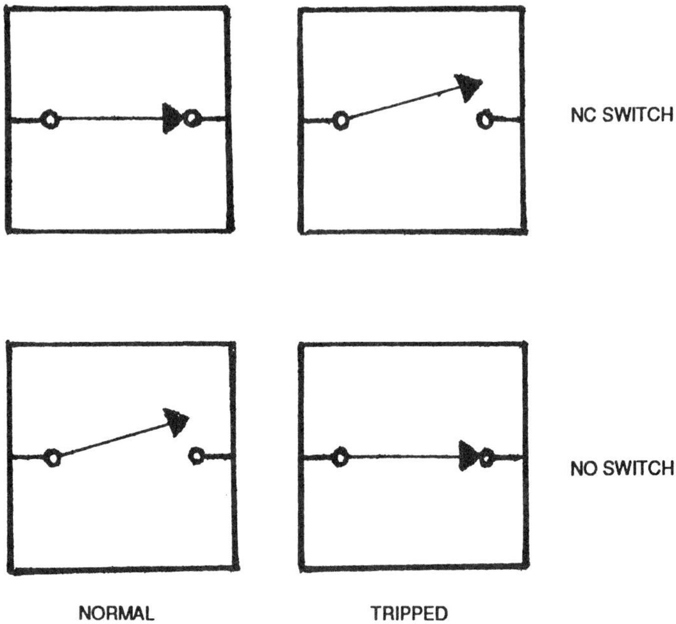

Figure 69. The difference between NO and NC switches.

es. The sensor switch will be either normally open or normally closed (fig. 69). A normally closed (NC) switch is a switch, magnetic or mechanical, in which the contacts are closed (electrically conductive) when no external forces act upon the switch. A normally open (NO) switch, on the other hand, is one in which the contacts are open or separated (electrically nonconductive) when no external forces act upon it. The control unit will sense when the status of one of its sensors is changing and then sound the alarm.

Therefore, a closed-circuit system is one in which the switches and sensors are connected in series. The alarm is sounded when an activated switch or sensor breaks the circuit or the connecting wire is cut. An open-circuit system, on the other hand, is one in which the switches are connected in parallel. The alarm is sounded when an activated switch closes the circuit, permitting current to flow through it. Open-circuit systems are not as secure as closed-circuit systems.

The alarm system must also contain some means for arming and disarming the system. Arming is the means by which an alarm system is switched on. It may be either manual, passive, or remote control. Disarming (i.e., switching off the system) is generally done in the same way.

Some self-contained systems are armed and disarmed by a key switch, while a smaller number are armed and disarmed through a code entered from a keyboard.

A passively armed system will arm and disarm itself when, for instance, the correct key is used for locking and unlocking the front door. This will be further described later in this chapter.

Many alarm systems, however, feature a delayed exit/entrance circuit that permits the user to leave and enter the premises without setting off the alarm. The time lapse is fixed, generally between twenty seconds and two minutes, and can usually be adjusted by the user.

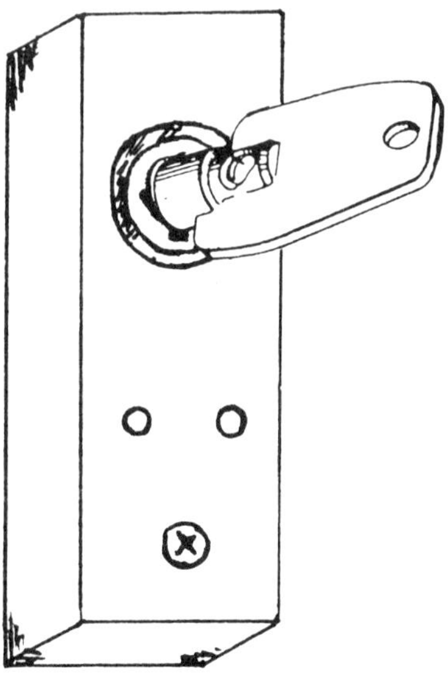

Figure 70. Key switch.

Another common option, especially in simpler alarm systems, is to arm and disarm the system with the help of a key switch (fig. 70). This seems to be especially true of alarm systems in ordinary American houses. The switch is mounted outside the house, and the alarm system will then not have a built-in delay. The lock used in such a system is difficult, but not impossible, to pick. It is generally a tubular cylinder lock. Furthermore, the interior of the lock panel has a tamper switch to prevent somebody from defeating the alarm. If the switch is damaged, the alarm will sound.

The tamper switch terminals can also be used as a panic button. A panic button is any switch connected to an instantaneously triggered loop. Panic buttons can generally be used whether the alarm is armed or not.

The tamper switch inside the control panel cover is not the alarm system's only protection against sabotage. A self-actuating warning device has an internal battery inside the sounder box. This will take over as power for the siren if an intruder cuts the wire. Tamper protection similar to that mentioned above will often be employed within the sounder box as well.

Other common features used to prevent a skilled intruder from disarming the alarm include contacts which are not easily visible on the surface of door or window frames, hidden components of the alarm system, and the use of four-core cable in the wiring. This cable will allow both closed-circuit devices, such as magnetic contacts, and open-circuit devices, such as pressure mats, to be connected within the same cable. With the former, the alarm will sound if the cable is cut but not if the cable is bridged. With open-circuit devices, cutting the connecting wires will not trigger the alarm, but bridging the circuit will. One pair of wires is used to form the closed circuit, while the other pair is used for the open circuit. The alarm will be set off by one or the other of the alarms if the cable is interfered with.

Even if one of these two types of sensors is not connected, many systems allow the other circuit to run into the sensor and then back to the control unit. This will protect the system in exactly the same way. It will be effectively impossible to sabotage the sensor.

Four-core cable, for the same reason, is also often used together with more advanced warning devices. If the warning device has an integral battery backup, it will then operate even if tampered with or if the wires connecting it to the control unit are cut.

Another option is known as line supervision. This is electronic protection of an alarm line accomplished by sending a continuous or coded signal through a circuit. A change in the circuit characteristics, such as a change in impedance due to the circuit having been tampered with, will be detected and initiate an alarm if the change exceeds a certain level. A normally closed loop is, in effect, a supervised line.

CONTROL UNITS

The control unit is the brain of the alarm system and fulfills several purposes. It is used to turn the alarm on and off, and it checks the circuits for faults. The control unit monitors the condition of all sensors and sounds the alarm by transmitting a signal to the warning devices whenever it detects a problem. The warning device will also be activated by the control unit if a sensor signals the unit. Of course, the control unit will only fulfill these duties when it is turned on.

The control unit, and in its extension the entire alarm system, functions in the following way. The control unit is connected to the alarm switches, which are connected in a loop circuit. The control unit includes circuits for checking the change of status in these switches, thus detecting when a switch opens or closes (depending on the type). This change will activate the alarm circuit. When the control unit has been activated by a switch, the alarm system is said to be tripped. This activates and sounds the alarm.

Loops of alarm switches are connected to the input terminals of the control unit. There are generally separate terminals for connecting a number of switches, or loops, to the control unit. Each terminal will identify a different loop of switches in the system. Remember that there will be a separate loop for every easily defined area in the building protected by the system. One loop might protect the front door, for instance, while another protects all windows, and so on. In this case, the front door loop will probably have a built-in delay, as this is the entrance to the house, and the owner must have some time to disarm the alarm system. The other loops of the perimeter defense will trigger instantaneously, as these areas are only used for entering and exiting the house by an intruder.

There might also be what is generally known as a day loop. This will also trigger instantaneously, but it will usually only activate a small buzzer instead of the main warning devices. The day loop is sometimes used by families with young children and warns if the children leave through the front or back door or open the gate leading to a swimming pool. This is not really a part of the intruder alarm system, although it uses the same circuits to a certain extent.

The alarm system is said to be armed, or set, whenever the control unit is in operation. Otherwise the alarm system is disarmed. Although control units with more complex circuitry also exist, this is the basic configuration of a control unit. Other features do not much affect the basic functioning of the alarm system. For instance, a system might turn off the warning device when the alarm has sounded for a certain period of time, often five minutes, and then reset the system. Other control units will assign priorities when signals are received from more than one loop in the sensor system. The loop used for fire or smoke detectors will generally override all other loops, for instance. Incidentally, this can be turned to the advantage of the intruder if the protected area is large enough. If a fire is started in one part of the complex during a break-in, the alarm might be effectively hindered from sounding in other parts of the building.

As this description demonstrates, a complex system will include several sensor loops. So the really critical connection is not between the control unit and the sensor loops, but between the control unit and the warning devices. If that line is cut, no warning will be sounded. The intruder should therefore attempt to break this connection rather than futilely trying to avoid the sensors, which is often almost impossible to do if the alarm system is designed properly.

The typical control unit will have both instant and delayed loops. An instant loop will sound the alarm immediately when a switch is tripped, while a delayed loop will wait for a certain period. The entire system will usually have a delay function, at least in modern systems. This allows sufficient time to arm the system on the way out and disarm it on the way in, and it obliterates the need for an external switch.

Most control units allow the different loops to be turned on or off separately. This is useful for guarding the entrance at night, even if the owner is sleeping in his bedroom, and for allowing the fire alarm to function at all times.

The number of sensors connected to one loop is not limited, even though there is only one set of terminals per loop. The switches are merely wired in series or in parallel (fig. 71). Normally closed switches are wired in series. When the switch opens, signifying that a window or a door is open, this breaks the NC actuating circuit in the control unit and the alarm sounds.

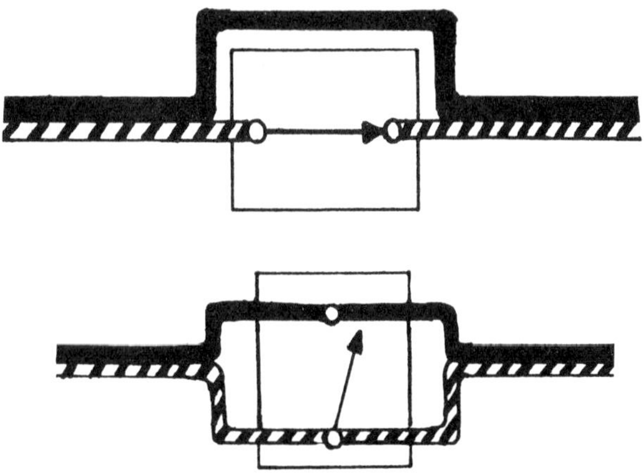

Figure 71. Wiring switches in series (top) and parallel (bottom).

Normally open switches, on the other hand, are wired in parallel. When any of the switches closes, signifying that an intruder has been detected, this completes the NO actuating circuit in the control unit and the alarm sounds. A normally closed switch can be bypassed by bridging the circuit. A normally open switch can be bypassed by cutting the circuit.

A loop of NC switches is generally more efficient, as it is being monitored continuously by the control unit. Any interruption of the NC loop will cause the control unit to sound the alarm. This is useful if the wiring connecting the switches is broken by an intruder, for instance. The NO loop can be cut easily, and then the control unit will not notice the difference. Remember, though, that a loop of NC switches is often run together with the NO loop in order to serve as a sabotage alarm.

In addition to this sabotage alarm, most professional control units include a plunger switch, known as a tamper switch. This will detect when the door of the control unit has been opened. As the tamper switch is on an instant loop, the alarm will sound as soon as an intruder tries to disable the system by damaging or turning off the control unit.

Often the control unit has an additional tamper switch installed in the back. This switch will sound the alarm if the control unit is removed from the wall. These two types of plunger switches are of the NC type.

As the control unit is often located in a cupboard, this door might also be fitted with an additional sensor that helps to guard the control unit, sounding the alarm when the door to the cupboard is opened.

The control unit can most often be found in an accessible but not easily seen place, such as a closet in the living room or hall, a bedroom closet, or in the kitchen pantry. It will most often be located in an accessible place, however, as it must be reached frequently to be armed or disarmed. An accessible location is not required if an exterior switch is used, of course. Instead of an exterior switch, an interior switch can be used, but this switch will most often be located in one of these accessible places, too, or else near the main entrance.

The control unit must always be installed in an insulated place, so there is generally no need to look in the attic, in a garage, or in the basement. A control unit must always be kept at room temperature. It is often firmly attached to the wall and cannot be removed easily without damaging both the wall and the control unit and also triggering the alarm. Sometimes the actual control unit is completely hidden, while the status of the system is shown on a separate control panel instead. This panel, even if tampered with, cannot control the system. It will only inform of the current status.

Modern alarm systems are often connected to the mortise lock in the front door. The system will be disarmed whenever the door lock is opened by the key. Picking the lock will produce the same result. One interesting way to disable certain control units is to remove the jumper wires between the unused NC connection terminals. This will, in effect, open the switch, and as the control unit senses an open circuit, it will produce an alarm. Such a minor tampering will leave the impression that the alarm system is faulty, and this might well cause it to be turned off until a maintenance man arrives. Of course, he will notice the real problem. This method requires that the operative disabling the system first gain entry to the control unit and then know how to distinguish between the various switches.

Warning Devices

Most alarms encountered will be local alarms (i.e., alarms that, when activated, will either make a loud noise at or near the protected area, flood the site with light, or both). There are also so-called remote alarms that transmit the alarm signal to a remote paging unit or monitoring station. These could be silent and not give any obvious local indication than an alarm has been transmitted. Sometimes a combination of local and remote alarms will be encountered.

Warning devices come in three basic shapes and in numerous types. All of them are suitable for installation in an alarm system, and any number or combination of warning devices can be used together. Most common are low-voltage sirens or bells, but regular Klaxons are also used fairly frequently.

The sirens are designed to emit a distinctive warbling high/low tone that can be heard over a long distance. In some alarm systems, a siren is used to indicate a burglary, while another siren or a mechanical bell, emitting a steady tone, is used to indicate a fire. An ordinary bell is, nowadays, seldom used as an intruder warning device. Instead of a bell, a buzzer can be used, but this, too, is a dated warning device, now generally used only indoors or in factories.

Today, the electric siren is absolutely the most common warning device used. A number of different types are used, but the piezoelectric siren is characterized by the lowest power requirements and is therefore often recommended. The warning device must usually be connected to a battery in case of power failure, so a low level of power is advantageous.

The siren in an intruder alarm system will invariably be of the rise-fall type, as is used in most European countries as a police siren. The design of the siren will not always be the same, however. Some of them are designed for outdoor use, while others are designed for indoor use only. The difference is not only in the level of protection against the weather that its cover will afford the interior circuits, but also in the noise these sirens will emit.

The aim of the outdoor siren is to be heard for as long a distance as possible, while that of the indoor siren is to cause discomfort to the intruder by emitting a really ear-splitting noise, thus scaring him away. For this reason, sirens emitting low-frequency sound are used outdoors, as they have the longest range, while high-frequency-sound sirens are used indoors. Their loud noise will serve as a very effective psychological deterrent to an intruder. Therefore, if the operative must perform a quick break-in without bothering to disarm the alarm system, it is worthwhile to use ear protection.

Yet another warning device, similar to a siren, is the pneumatic warning device. It relies on compressed air and therefore does not need much power. The compressed air, controlled by an electrical air regulator, will produce a very powerful sound when released. Most types of sirens will produce a sound of at least 100 decibels or more, and sound levels of up to 136 decibels are not uncommon.

Another type of warning device is the so-called silent alarm. This is an automatic telephone dialer that will place an emergency call to the owner, the local authorities, or someone else, and then play a taped message indicating the problem, whether it is a fire or a break-in. Telephone automatic dialers and other remote signaling systems will be detailed in the next section of this chapter.

Some alarm systems, especially those without a remote key switch, will also use a small prealarm buzzer or piezo sounder, which will sound during the entry delay time to remind the owner to disarm his alarm system before the real alarm is sounded.

Finally, there are strobe lights. This type of warning device is especially useful in a crowded neighborhood, where it might be difficult to pinpoint the origin of a siren. The strobe light, if installed, is often the weak link in an alarm system. Such a light will be connected in parallel with the siren or bell used as a warning device, as it is generally only an auxiliary system designed to scare away the intruder and help pinpoint the source of the warning sound from the siren. As this connects the strobe light electrically to the same contacts on the control unit as the siren, the entire alarm will be rendered useless if an intruder first inconspicuously short-circuits the strobe light. This can be done with water if the strobe light is not sufficiently weatherproof or equipped with a tamper switch.

Of course, the wire can also be cut. For this reason, the wiring will usually be hidden or at least out of reach. Despite this, it is remarkable that so many alarm systems can be disabled by simply cutting the wires to the siren or to another warning device.

The warning device, whether a siren or a strobe light, will often be mounted near the eaves or the roof line of the building, at a position at least 2.7 meters high. It might include a tamper switch and will generally be wired through the attic. Needless to say, a favorite tactic of professional burglars is to first of all break into the unprotected attic to locate and cut the wiring to the siren and then break into the house. Sometimes the warning device is instead mounted on a lumber or a metal pole, such as a television antenna. It is not uncommon to mount it directly on the existing mast of the television antenna. This is especially true of strobe lights.

Many, but not all, warning devices will come with a tamper switch, allowing the alarm to sound even if the siren is tampered with. Of course, this is worthless if no other warning device is present. Always look out for an extra hidden warning device before you disable the main one.

In most countries, the sounder box will turn off automatically within a preset time, usually no longer than twenty minutes, so as not to disturb the neighbors. The system will then reset itself automatically. This might not always be the case with the alarm system in a government installation, however. Even if the noise is turned off, there is often an affixed strobe light that will keep flashing until someone manually resets the alarm.

Remote Signaling Systems and Automatic Dialers

Nowadays, many alarm systems are linked to a telephone. By using a separate, exdirectory telephone line used only for outgoing calls, they are designed to raise the alarm in a remote location in case of intrusion. These devices, known alternatively as automated dialing equipment (ADE), automatic dialers, autodialers, or telephone dialers, are available in three slightly different varieties.

The reason for using an exdirectory line is that the alarm cannot then be neutralized by calling the automatic dialer. Placing such a call will effectively block the outgoing call. In some countries, the telephone system allows all incoming calls to be routed to a different number, thus freeing the automatic dialer's line.

The simplest type of automatic dialer is programmed to dial the local police and then play a standard prerecorded message stating that there is an intruder in the house. Of course, the message will also include relevant details, such as the owner's name, the address, and the telephone number. This system, although reliable in theory, is frequently useless, as many police forces no longer have the manpower to monitor the lines. Another drawback is that the automatic dialer will not work if the intruder cuts or temporarily disconnects the telephone lines. Furthermore, the automatic dialer will generally only dial its call once. If nobody answers, then it is bad luck for the owner of the system.

Another version of this system works in the same way, but instead of calling the police it will alert a neighbor or a relative who will then call the police. This system is even worse, as the friend might not be at home, and, even if he is home, the call will take still longer to reach the police. Here, again, a cut line will effectively prevent the system from sounding the alarm.

For these reasons, the digital communicator is more popular. This is a more sophisticated system that is able to dial a central monitoring station, usually the security company's central control station or the main security station in a corporate complex. Here, personnel will be on constant duty to observe annunciators reporting on the condition of the alarm system. The alarm call will consist of a series of coded signals that comes up as text on a computer screen which is monitored twenty-four hours a day by the staff. Upon observing these signals on the screen, a staff member will alert the police immediately and often dispatch a corporate security team to the location as well.

The digital communicator will continue to call until the message gets through. This is determined by the receiving station giving a correct code, which means that the message has been understood. In this case, the digital communicator might be programmed to call several numbers until it receives a confirmation from one of them. Some systems are designed instead to confirm by calling back within a specified time. Generally, this type of system is also designed to register any faults on the telephone line, thus announcing the danger of a cut line. This system, too, relies on using an exdirectory line.

Locations guarded by such systems are generally easy to recognize, as the company that installed them will advertise their presence with posted signs. Systems of this kind are almost always rented as parts of professionally installed alarm systems and are subject to regular maintenance inspections by the security company.

An even more advanced system, marketed in Britain and

in certain other countries, also responds to a fire alarm. It will furthermore indicate exactly which zone or loop the alarm has been triggered in and what type it is. Any faults on either the line or in the alarm system will also be reported.

The most exclusive alarm systems use direct private lines rather than the ordinary telephone lines. In this case, the alarm signal will be transmitted on a continuously monitored private line to the security company's central monitoring station. Any fault or interference with the line will be noticed. Such a system is in use only in high-risk installations or on the premises of the extremely wealthy.

If an ex-directory or a private line has not been used, the automatic dialer must always be connected as the primary telephone in the house. This means that all incoming calls must pass through the automatic dialer and then proceed to the regular telephone. If this is not the case, the system, even if it is hidden, can be easily disabled by removing the regular telephone's handset from the hook.

Some automatic dialers have their own built-in backup batteries, which allow them to work for several hours even if the power is cut. The power will ordinarily come either from the control unit or from a separate AC power adapter.

The recorded message will either be on a magnetic tape, in which case the owner can record his own message, or it will be a prerecorded computerized voice. Most automatic dialers can be programmed with up to three telephone numbers, all of which will be called in order.

An interesting point about some of these systems, especially the British ones, is that some police forces insist that the siren, if one is used in conjunction with the remote signaling device, have a built-in delay so that it will sound the alarm three to five minutes after the message has been relayed to the police. This is to give the police a greater chance of the catching the intruder in the act.

It should also be noted that even if the automatic dialer is connected to a security company, it is by no means certain that they will care to respond to a single indication of a triggered alarm. False alarms are now so common that most companies of this type will wait until they first have received an alarm from an external detector, then from a perimeter detector, and finally from an interior detector, indicating that this is a serious intrusion attempt. Only then will they dispatch a patrol or call the police. This reluctance to respond to false alarms will give the operative a few valuable minutes in which to execute his mission and get away.

A radio transmitter can also be used as an automated dialing system. In this case, it is most common to use a transmitter in the frequency-range around 27 MHz, as this is most commonly used in personal paging systems. A radio system utilized in this way is most common in advanced alarm systems in vehicles, boats, and other places in which there is no regular telephone connection.

The suitability of the antenna is the most important factor, as this will determine the range of the system. If the antenna is removed, or covered by a metal box, the transmission will suffer a severely decreased range or even disappear completely. Naturally, no remote alarm will then be sounded.

Access and Exit Control Systems

Access control is the means by which only authorized persons are allowed to enter a building or flat while unauthorized persons are kept out. Such a system is commonly designed around an exit-entry control system and/or a system to arm and disarm the alarm system. For these purposes, there will be an authorized access switch that makes all or parts of an alarm system inoperative in order to permit authorized access.

In most doors the mechanical lock is the only access control system. There are, however, a large number of other possibilities, mechanical and electrical, sometimes used only as keyless locking devices, but sometimes also used together with alarm systems. In the latter case, the system will definitely rely on electrical control switches. Most of these control systems require the use of combination codes. See Chapter 2 for more information on this.

Electrically operated locks are becoming more and more popular, especially in industrial installations and offices. So far, however, electrically actuated release catches are more common than pure electronic locks. These units operate on low voltage, often 24 V. For this reason, they need transformers. Locks that are normally locked will be unlocked when the system is energized. On the other hand, those that are normally unlocked will lock when the unit is energized. The first option is of course most commonly encountered by an operative desiring to enter.

It is sometimes possible to enter by connecting a high power source to the lock, thus overloading the circuit. However, there is also another factor that must be taken into account. Some locks of this type are so called fail-safe. This means that they will automatically unlock if the power fails, such as might happen in an emergency such as a fire. This is to provide a safe escape route, of course, but it can serve equally well for purposes of gaining entry to the premises. In short, tampering with the power supply might well open a lock of this type, as long as you know what type of locking device you are dealing with.

Also note that there might be a considerable distance between the actual lock and the remote control unit. Most commonly, the control unit is next to the door, and it is always built around a control switch. There are many types of control switches. Among them are digital access control systems, electronic or mechanical card access control systems, lock switches, remote control switches, delayed alarms, and ordinary key switches. The various systems may rely on number code combinations, coded cards, or plastic keys instead of metal keys. This is common in hotels and in many hospitals, for instance. As these systems are quite often electronic, they are also frequently connected to a registration unit, able to record when a certain code or key is used and where, if the system includes more than one lock.

The digital access control system is perhaps the most common of these systems. It usually consists of two parts: an access control keyboard and a program unit. Such a system can possibly use ten thousand different code combinations. Every legitimate user might have an individual code, or everyone might use the same one. The code might be used for opening the lock or for disarming or arming the alarm system.

Less commonly, a dial of the type used in combination locks can also be used in this device. Both types are very popular with large companies, as it might be necessary to change the code from time to time, such as when staff is replaced.

There are digital systems that only close the circuit momentarily, thus turning off the alarm system and/or opening the door for up to ten seconds or so. Others remain closed until the code is reentered. Some systems allow the use of different codes for different individuals. Most of these switches will only disarm one loop, consisting of the sensors that guard the nearest way to the control unit, so that the owner can proceed there immediately to disarm the entire system.

Usually the code will consist of either four or six digits. Occasionally the four-digit code will serve as a code lock, while the six digits will arm or disarm the alarm system. Sometimes the lock will not unlock until the alarm system has been disarmed. It is also common for the code lock to temporarily block another attempt if the code entered is the wrong one. This is to discourage attempts to enter by using random combinations. Some systems go even further, triggering an alarm when a preset number of incorrect combinations has been entered.

Most systems of this type will also have a preset timer that is activated when the first digit of the code combination is entered. If the remainder of the code is not entered before the time expires, the entry is canceled.

Yet another interesting feature of some of these systems is what is known as a duress alarm. This is connected to an alarm system and a warning device, such as a siren. A silent alarm is also a real possibility here. The duress alarm is activated by depressing the correct combination code but replacing the last digit with a predetermined other digit, known as the duress digit. This will or will not open the lock, depending on the programming, but it will definitely trigger the alarm. This is to warn against entry made under duress, for instance, by an employee who is held at gunpoint by an intruder. Technically, this system is a momentary switch with NO switch contacts.

Another unit, very often used in conjunction with this device, is the card reader. A card-based access control system is easy to use and therefore popular in large corporations and government installations. The card reader is situated next to the door, requiring the person wishing to enter to both insert his coded access card in the reader and punch his personal code on the keyboard.

The plastic cards for these locks are of two different types. They can either rely on a magnetic code, in the same way as a credit card, or they can have various optically read numbers or figures. In either case, the card must be entered into a slot or passed through a card reader situated at the door. There are also systems in which the card reader is invisible, hidden in the wall next to the door, and the card is simply displayed roughly 30 centimeters from the reader in the wall.

Whatever type is used, the card will have a code that allows entry through only one or several doors, depending on the design and programming. All cards can have the same code, or individual codes might be used instead. The system is extremely flexible. A plastic key with a code in it can also be used. This works exactly like the card keys but is designed to be carried like an ordinary key on a key ring.

In certain of these locks, however, the system is not electronic but mechanical, even though the key is still replaced by a plastic card. When a plastic card with holes in it is used, the lock is definitely mechanical. The holes in the card will fit exactly to a number of balls within the lock. The lock can be individually coded, and the code can be changed easily. The card might have a code allowing access through several doors or only one. Note that this is not a real code, but only a means of ensuring that the card fits into its slot in order to open the door. Consequently, it is the same as the cuts on a regular key.

A door protected by a code lock or card lock will often have an alarm sensor as well, which will sound the alarm if the door remains open too long. Sometimes this will only be a buzzer to indicate that the door must be closed, but occasionally a real warning device will be used.

Electronic locks may also take advantage of time coding. This is a system that is programmed to allow or deny access, depending on the time of the day or night. Every legitimate user might be allowed to enter during ordinary office hours, but only key personnel will be permitted to enter after office hours. When a system of this complexity is used, an automatic registration unit will almost certainly exist. A system of this type can be made very flexible indeed.

If a card is lost, the code (if one is being used) will be changed as soon as the owner becomes aware of the loss. The person who lost the card will simply be issued a new card and a new code. Changing the code is a quick process. The number of possible combinations is very high.

It is difficult to bypass access control systems of this type. The keyboards and card readers might be vulnerable to weather, but this will not allow access; it will only prevent it further. There are, however, a few ways of cheating such a system.

If the same code has been used for a very long time, it is sometimes possible to see which numbers are being used, as wear and tear together with dirt will show which keys are used most commonly. The combination is then not obvious, of course , but the number of possible combinations will decrease significantly. This has often helped the enterprising operative to gain entrance.

Another common way of learning the code, at least in those locations where no card is required, is to simply observe somebody entering from a distance. However, many keyboards have been fitted with protective covers to prevent this from happening.

Electric digital code switches can sometimes be shorted by connecting a high-power cable to them instead of the ordinary low-power cable normally used. This might open the lock, but only if it is of the correct type, the one that will unlock when energized.

In private homes, the access control is usually simpler.

One common access control system is the lock switch. This device fulfills two purposes. The first is to disarm the alarm system, while the second is to unlock the door. The system is usually installed in the mortise lock in the front door (see the section of this chapter on lock switches built into standard, mechanical mortise locks). This is common in private homes, where the number of keys is limited. When this system is used, the door and its frame will often be protected by an inertia sensor, for instance (see the section on vibration detectors and inertia sensors in Chapter 7).

Remote control switches are another possibility. They come in two major types. The most common is the radio transmitter operated switch, but infrared operated switches are also used commonly. This system is often used in garage doors. The remote control will open the door and disarm the alarm system at the same time. Remote control switches, especially of the radio frequency (RF) type, are also common in perimeter alarm systems around outlying sheds and stores. The alarm is then built around the sensors mounted on the surrounding fence, but the remote control device can arm and disarm the control unit, located inside the shed or store. The range will then be around 70 meters, which is usually enough.

The radio transmitter will transmit a digitally coded signal. This signal, when recognized by the receiver switch, will arm or disarm the alarm system, or parts of it, if so desired.

Delayed alarm systems are also common. The alarm will simply be delayed so that the operator has enough time to reach the control unit and turn off the system before the alarm is sounded. The delayed alarm was described in more detail in the first section of this chapter.

It should be pointed out that some corporate alarm systems are not armed and disarmed manually at all. Instead they rely on a timer switch that will turn the alarm system on and off before and after office hours. Note that the time will not necessarily be the same every day of the week.

A key switch, finally, is exactly the same as a standard lock, but it is connected to the alarm system. It is usually in a small box, protected against sabotage by a tamper switch, and located outside the building. As the switch can sometimes be tampered with by exposing the wires going to the key switch, it is often at least set into hard material and is usually also protected by an inertia sensor.

Electric locks were described previously, but another problem might be electromagnetic locks. These are used in hospitals, banks, prisons, airports, and numerous other high-risk installations. The interesting thing about the electromagnetic lock is that it can exert a very strong holding force. In fact, the lock consists of two components, the lock itself and its armature (fig. 72). The lock is mounted to the door frame, while the armature is mounted to the door. Both mountings are designed to be sturdy and resistant to physical abuse. The lock and the armature will make con-

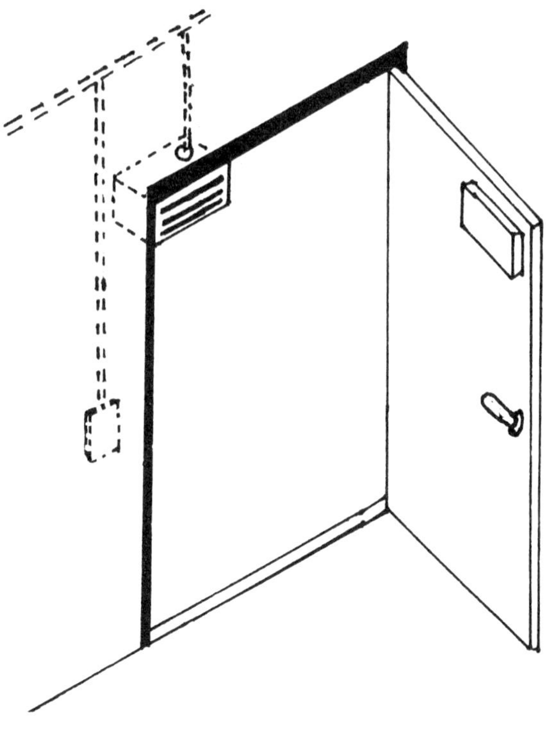

Figure 72. Electromagnetic lock.

tact when the door is closed. Locking, or activating, the lock will cause the two units to be magnetically attracted to each other and hold together. The wiring will be factory-made and includes tamper-resistant circuits. The easiest way to disable the electromagnetic lock is to deprive it of energy. However, this is not always so easy.

Lock Switches Built into Standard Mechanical Mortise Locks

A lock switch can also be built into a standard mechanical mortise lock (fig. 73). The lock switch fulfills two purposes. The first is to disarm the alarm system, while the second is to unlock the door. The standard key to the lock is used, and while the key will unlock the lock, the switch will disarm the alarm system. The system is usually installed in the mortise lock in the front door, although other secure locks, such as tubular locks, might also be used for this purpose.

This device is common in private homes, where the number of keys is limited. When this system is used, the door and its frame will often be protected by an inertia sensor. It is possible to pick this lock. An even greater danger is the key to the lock, as this will not only allow

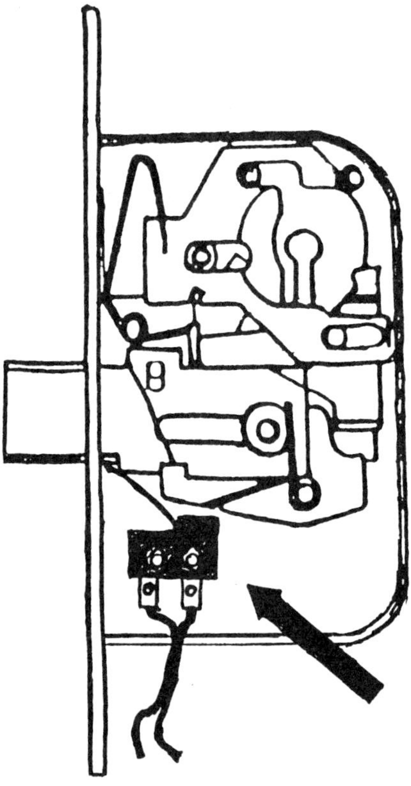

Figure 73. Lock switch installed in mortise lock.

entry to the building but disarm the entire alarm system as well.

The lock switch is a standard contact switch, which is mounted inside the lock next to the latch bolt. The switch will react when the latch bolt is moved to the locked or unlocked position. Such a switch is, in itself, relatively easy to manipulate, so other means of protecting the switch are usually installed.

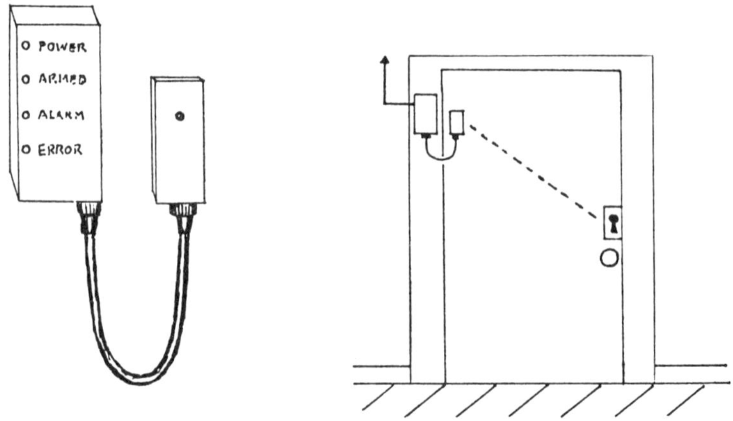

Figure 74. The jumper cable terminal and its position on the door frame.

One efficient protective system is to use a mortise lock along with an inertia sensor built into one of the jumper cable terminals mounted on the door frame (fig. 74). This sensor will protect both the lock and the door. There is also an LED in the face of the lock. This LED will not only indicate that the alarm is armed by flashing, it will also work as a sabotage protection, as its circuit will trigger the alarm if it or the lock is damaged. Furthermore, if the power is not sufficient, the LED will stop flashing and instead remain on continuously. This also happens if the alarm has been triggered. The system will reset automatically every time the door is locked.

The switch circuit is a NO system, open when the lock is locked and closed when the lock is unlocked. The wire is usually of the four-conductor type, however, two of them being the NO circuit and the other two the NC sabotage defense. The wiring is run either through a diagonal hole in the door or on top of the door. An alarm of this type can also be used if several doors are to be used as main entrances. The locks and their systems will then be connected in parallel. Then the alarm system will be disarmed whenever any of the locks are unlocked. However, the system will be armed only when the last unlocked lock is finally locked.

POWER SUPPLY AND BATTERIES

Most alarm systems are powered through the control unit. Generally, the power supply is housed in a separate transformer box plugged directly into a wall outlet. The transformer converts the AC current to 12 volts DC, which is used for the entire alarm system. The transformer is attached to the control unit by an ordinary two-conductor wire.

If the intruder can legitimately gain entrance to the building to be searched, for instance during office hours, it is sometimes possible to locate the transformer and simply remove it from the wall outlet. As most alarm systems will have a battery backup power source, the alarm system will switch to battery power at once, usually without alerting the inhabitants of the house. This is generally enough to last for several hours, but it will run out soon after the occupants have left the building in the evening or gone to bed at night. Such a method is often helped by the fact that most wall outlets are in low and inconspicuous locations, such as behind furniture. Even though the control box might be checked in the evening, the transformer will most often not be checked.

Backup batteries are commonly used because burglars have learned to shut off the electricity in the house they are going to enter. The backup batteries are rechargeable and are usually installed in the control unit. The rate of discharge depends mainly on whether a siren has been activated or not, as this consumes a large amount of power, but the backup protection will generally last for around three hours—or less if the siren is used.

It should be noted, however, that in many European countries the backup batteries are more often designed to last for at least twenty-four hours or even seventy-two hours (a normal weekend). If the alarm system has no backup batteries, the system will always come back on armed if the AC power has been cut off temporarily.

Some control units will check the status of the batteries each time the system is armed or disarmed. This is to inform the user of the fact that power is low.

As a curiosity, it might be mentioned that certain external control panels, installed to monitor the current status of the alarm system, also monitor the fact that the current required to recharge the backup batteries is insufficient. This is to advise the owner that he must check and possibly repair this function. Opportunist burglars are thankful for this information, as they then know that it is safe to break into the house as long as they cut the power first.

Wiring

Most wiring connecting the various sensors to the control unit will be concealed, or at least located very unobtrusively. The wiring is frequently run inside the interior walls. A small hole will be drilled near the sensor. The wiring will go through this hole into a larger hole in the sole or top plate in the basement or the attic. The wiring will be run there, following the attic or basement until it once again penetrates into and surfaces through a small hole next to the control unit.

Wiring can also be hidden under carpeting, beneath floorboards and masonry trim, or behind furniture. It is generally not placed where it might get damaged by water, excessive moisture, or local pests (rats and other rodents tend to bite through electrical wiring).

It was mentioned above that a loop of NC switches is often run together with the NO loop in a four-strand cable to serve as a sabotage alarm. This is not the only way of protecting the alarm system against sabotage. The most inconspicuous defense is the balanced alarm system. This is impossible to detect from the outside or even from the inside of the building, as long as the control unit is not found and opened.

A balanced system consists of a number of sensors, just as an ordinary system, but every sensor will be fitted with a unique resistor. The control unit will then be adjusted to recognize the total resistance of the system. If one sensor is shorted or bypassed, the control unit will detect a decrease in the system's total resistance. This will trigger the alarm immediately.

It should also be remembered that some contemporary alarm systems utilize built-in computers to control the system. This means, among other things, that any changes in the system's status will be recorded, even if the alarm has not been sounded.

More advanced alarm systems will employ computerized communication on two or four wires to further protect the system against sabotage. The computer will regularly query the various detectors about their status. The entire net of wiring will be monitored in the same way. As this system is almost always connected to a central alarm, the entire system, including any detected irregularities as well as triggered alarms, can be presented on a computer monitor. One example of this system is the U.S. Vindicator system, installed to protect air bases and important defense industries but now commercially available to major companies as well.

In this case, every detector is connected to a transponder, which will report through time-multiplexing every change in wiring or any indication given by the detector. These reports will be presented in real time to an alarm operator in the security central. The system is therefore extremely difficult to bypass, and here it is definitely easier to try to subvert the human computer operator in the system, as he is the weakest link in the chain.

Wireless Systems

Wireless alarm systems are also used fairly frequently.

They are more expensive, but they are also easier to install than the wired systems. Such an alarm system consists of a central processor unit and one or more detection devices. All these components communicate by using radio waves. The radio frequencies used are supposedly free of interference from any other radio-controlled equipment such as the communication radios in passing taxis and police cars.

The entire alarm system is controlled by a small radio touch-pad, about the size of a pocket calculator. This remote control can be used to activate the alarm from anywhere in the house. It can also be used as a personal attack alarm, or panic button, wherever the user is, whether in the house or outside in an adjacent garden or garage.

In the United States, some wireless alarm systems are available that not only sound the alarm but also regulate the turning on and off of lights and other electric appliances, such as television sets. These devices are supposed to scare away intruders by pretending that somebody is in the house, even when it is in fact empty. This device is either battery-powered, or it takes its power directly from the AC power supply. In the latter case, the entire system will fail if the power is cut. If batteries are being used instead, the battery-powered transmitters will be supervised every ninety minutes or so to determine battery and functional condition. This will safeguard against loss of power, but only as long as the control unit itself remains powered, of course.

Wireless alarm systems must not be mixed up with self-contained alarm systems. The latter also sometimes rely on radio, but work in a different way. A real wireless alarm system is in effect a separate-components system, but using radio waves instead of wiring.

CHAPTER 7

Alarm Sensors

There are many different types of sensors, and they are often categorized broadly as active and passive. Active sensors create a field and detect a disturbance in that field, while passive sensors detect natural radiation or radiation disturbances without themselves emitting the radiation on which the sensor's operation depends.

Sensors are also divided into categories based on the area or particular point that they protect. Perimeter protection prevents access to the outer limits of a protected area by means of physical barriers, sensors on these barriers, or external sensors not associated with any physical barrier. Interior protection is a line of protection along the interior boundary of a protected area, usually a building, including all points through which entry can be made. Area protection, finally, covers an inner space or volume of a secured area by means of a volumetric sensor, or a sensor with a detection zone that extends over a volume, such as an entire room.

Magnetic Reed Switches and Wire Contact Systems

A magnetic reed switch is an alarm system detection device in which a disruption of the magnetic field between two points causes a break in the electrical current. This

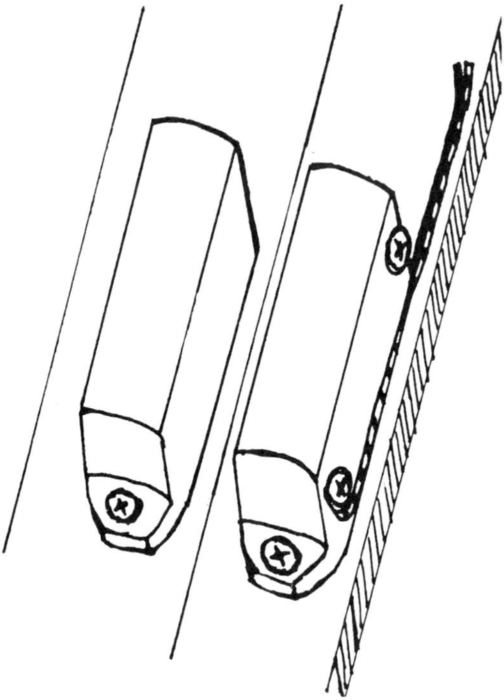

Figure 75. Magnetic reed switch.

break signals the control unit to activate the alarm. Magnetic reed switches (fig. 75) are among the most common alarm sensors in use today.

Magnetic reed switches are a type of magnetic switch that consists of electrical contacts formed by two thin, magnetically actuated, reedlike metal vanes, held in position (normally open or normally closed) within a sealed glass tube. The tube is enclosed in a metal or plastic case. The device works on the principle of magnetic attraction. The sealed metal contacts are positioned in such a way that when a sufficiently strong magnetic field is present, they are either pulled together or pushed apart (fig. 76). When the magnetic field is removed, the metal contacts will naturally move in opposite directions.

The reed switch is therefore composed of two separate units: the magnetically actuated switch and a large magnet, which is enclosed in a similar plastic housing.

The contacts within an NO reed switch do not touch

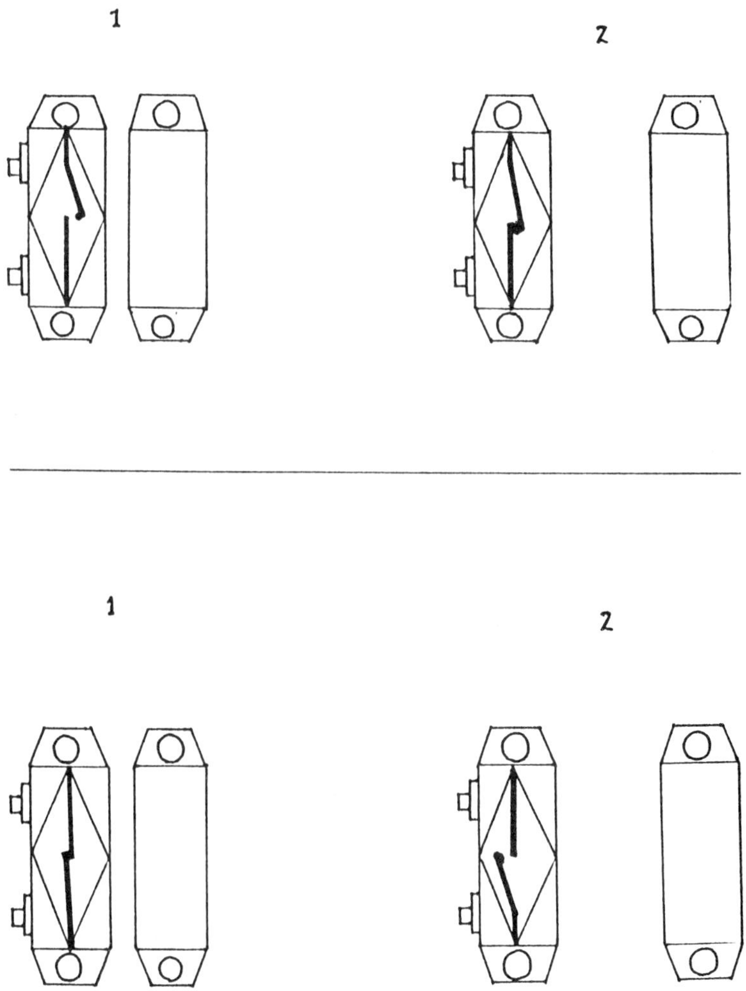

Figure 76. NO and NC magnetic reed switches.

when the magnetic field is strong enough. The loop will be normally open. But when the magnet is removed, the contacts will be pulled together, closing the contacts. Of course, the opposite is true of the NC reed switch, in which the contacts are closed when the magnetic field is affecting them. If the magnet is removed, the contacts will open.

The magnetic reed switches are set into the doors and windows in each zone, either surface mounted or recessed. Recessed switches (fig. 77) are invisible when the door or window is closed, being set into recesses in the frame. For this reason, they are not as easily tampered with as the surface-mounted reed switches. The circuit will be broken whenever the door is opened, and this will trigger the alarm.

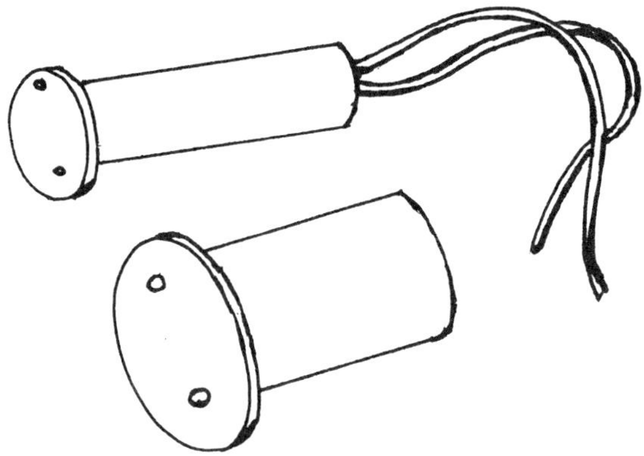

Figure 77. Recessed magnetic reed switch.

The switch is usually mounted in a fixed position, such as a window frame or a door jamb, opposite the magnet, while the magnet is fastened to the window or door. When the window or door is opened, the removal of the magnet will force the switch contacts to change status. This will open or close them, depending on the type of switch. In either case, the alarm will sound.

NC reed switches are the most common, as they are most secure. Of course, NO reed switches can also be used, but, once again, they can be circumvented by simply cutting the wiring.

Magnetic reed switches will be found as far away as possible from the hinges, as they then will trigger the alarm even if the door or window is only partially opened. The switch will generally only be located near the hinges if the owner wants to open the window at night for ventilation. An ordinary reed switch generally triggers when the magnet is moved away more than about 2 or 3 centimeters, although some switches, so-called wide gap switches, allow the window to be opened 5 centimeters or so. The latter switches are used when the construction of the window or door does not allow the two pieces to be mounted closely together. They are also used if the owner of the system wishes to keep a window slightly open for ventilation.

Reed switches come in different shapes and brands, but they are typically about 3 1/2 to 4 centimeters long and a little more than a centimeter thick and wide. They are, for obvious reasons, always mounted in pairs, one of the pieces (always the one on the moving part of the door or window) housing the magnet and the other housing the stationary switch part mounted on the door jamb or window sill. When the window or door is closed, the two parts will be very close to each other. Look out for the wiring, which is sometimes hidden in the wood. The terminal screws for the wiring will generally be visible, however.

Some alarm systems also include miniature reed switches. They work in exactly the same way as ordinary reed switches but must be aligned properly and placed very close to each other, as the magnet is smaller and the magnetic field consequently not as strong.

Sometimes reed switches can be found in other locations, protecting, for instance, cabinet and cupboard doors, internal doors, or even garden gates.

It should be mentioned that some self-contained window/door alarm systems also rely on magnetic reed switches. The only difference is that the switch section is fully integrated with the control unit. Some of these devices can even double as door chimes when disarmed.

Magnetic reed switches are generally located in a separate loop, as they will only sound the alarm once if the intruder leaves the door or window open after entering. If they are positioned in a loop together with other sensors, this loop can easily be nullified by leaving the reed switch and its magnet well apart from each other. No other sensor

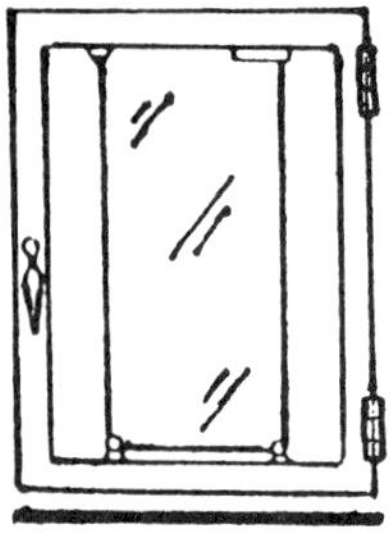

Figure 78. Wire contacts.

can then sound the alarm after it has been sounded once by the reed switch.

Wire contact systems are now very old-fashioned, but they can still be found occasionally. The wire contact switch relies upon a wire or thread that is connected both to the switch and to another, fixed point. The switch will sound the alarm if the thread is either put under tension or slackened. As the thread is generally an electrically conductive wire, the wire will also function as a sabotage alarm. Therefore, the alarm will sound if the wire is cut or broken. Wire systems of this kind are generally used to protect windows and walls (fig. 78) or serve as external alarms in gardens, for instance.

Window Foil

Adhesive window foil, although nowadays clearly dated as an alarm sensor, is still in common use, especially in shops. It is not commonly used in residential alarm systems. Window foil looks like silvery lead foil and can be found around the edges of the window (fig. 79). The foil consists of thin metallic strips made of a lead-aluminum alloy, which are cemented to the protected surface, usually glass in a door or window. The metallic strips are connected to a closed electrical circuit. If the protected material is broken, and the foil as well, the circuit opens. This will initiate the alarm. Sometimes foil is simply called tape.

Figure 79. Window foil.

Window foil has many disadvantages. The main one is that a foil alarm is a one-time alarm only. When the window has been broken and the alarm has sounded once, anybody is free to enter until the window and the circuit have been repaired. Another disadvantage is that the foil deteriorates with age and often becomes too brittle to function after a few years. Yet another disadvantage is that the foil will not sound the alarm if the intruder simply cuts the window open with a diamond, without breaking the foil. Then repair or replacement is necessary. Finally, windows protected by foil generally cannot be opened. For these reasons, other alarm sensors, such as glass breakage detectors, are better investments and are therefore used more commonly.

If the foil is installed on a window that needs to be opened occasionally, it will be connected to a contact strip, which is used to disconnect the wires from the window foil block. The spring section is mounted on the window itself, while the contact plate is on the window sill. The metal tabs on both sides of the switch must make good contact.

The connection between the window foil and the control unit is, under normal circumstances, active only when the window is closed. Window foil is used with the NC circuits of the system. It must therefore be disconnected if the window is to be opened. If so, the switch must be prepared for a coiled jumper wire that will temporarily bridge the gap between the two parts of the contact strip.

There have been attempts to make more reliable versions of window foil. Certain manufacturers of insulated windows include a very narrow and thin metal strip in the glass, usually hidden by the rubber strip used for insulating the window. They reason that the strip will break if the window is broken. This alarm system is more safe, as the metal strip is effectively invisible and securely located, protected by the glass. However, this is also a one-time-only alarm and suffers most of the same disadvantages as ordinary window foil.

Windowpane-Mounted Glass Breakage Detectors

Glass breakage detectors (fig. 80) come in several different types, but they all are designed to detect when a window is broken or otherwise removed from its frame. In either case, the sensor is attached to the glass on the inside of the win-

Figure 81. Vibration detector.

dow. The sensor is fastened with double-sided adhesive tape, but it can be removed easily if required. There are versions available both as NC and NO switches. One detector per window is generally sufficient, unless the windows are very large, such as in major offices or shops.

So far, the three most common types of glass breakage detectors include the mercury switch type, the weighted arm type, and the electronic or "tuning fork" type.

The mercury switch type, obviously, relies on mercury. At room temperature, mercury is a liquid metal that conducts electricity. The detector is designed to have a small pool of mercury near an NO switch. If the glass in the window on which the detector is mounted is broken or severely shaken, the switch will be tilted or moved and the mercury will come into contact with the switch terminals. This will close the switch and trigger the alarm.

The weighted arm type uses a movable internal contact that is weighted so it remains in one position. However, most vibrations will make the arm move. This will either complete or break an internal switch contact, depending on how it is designed. The contact will trigger the alarm.

The electronic glass breakage detector uses an internal tuning fork which vibrates when the window is broken or jimmied. The vibrating tuning fork triggers the alarm.

Glass breakage detectors of these types, in the same way as window foil, can use a contact strip if the window needs to be opened sometimes. All of them share the same problem, however. They are susceptible to any vibrations in the window, including those caused by the wind. This might produce numerous false alarms, especially if the window is not installed in its frame tightly. For this reason, other types of glass breakage detectors have been introduced that work by detecting the sound frequencies emitted by breaking float glass and/or shattered window frames.

This type of detector can detect and identify the special noise that is heard when glass is smashed, when a diamond is used for cutting glass, or when metal hits glass. For this purpose, the detector incorporates a piezoelectric microphone with a resonating frequency within the range emitted when glass is broken. Another option is a microphone sensitive to frequencies above 60 kHz, so that the glass breakage detector is not sensitive to noise in lower frequencies, such as that which results from heavy traffic. The detector is generally fixed to the glass it is protecting, although some can also be put on adjacent walls. The latter type will be detailed in the next section of this chapter.

The kind of glass is important, as the majority of these glass breakage detectors will only react to that of breaking float glass, but not to the sounds of breaking laminated, tempered, or wired glass. Some of these detectors are also susceptible to false alarms, being set off by the sound frequencies of rattling keys or bottles, for instance.

Vibration Detectors and Inertia Sensors

A vibration detector is one which is placed on walls or window frames to register vibration caused by blows, drilling, or breaking glass (fig. 81). Self-contained units, to be fixed to doors, are also available. The sensor will signal an alarm whenever it registers the kind of vibrations that it is programmed to identify as signifying an attack. Vibration detectors, or shock sensors, come in different types, some of them relying on mechanical means and others working on electronic principles.

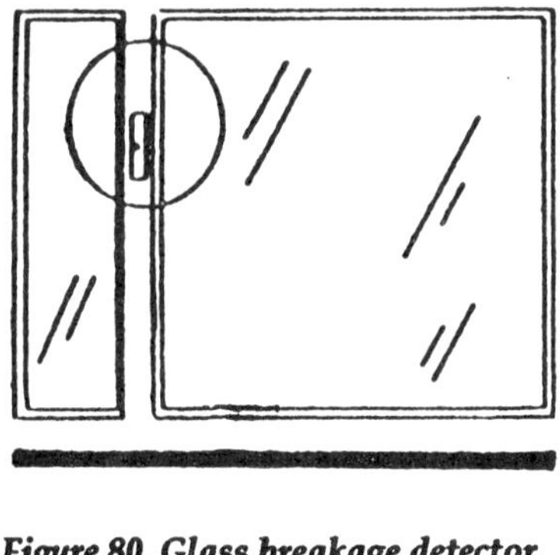

Figure 80. Glass breakage detector.

The most common vibration detector is still the pendulum alarm. This is a mechanical detector that relies on a pendulum switch, used to sense vibration or motion. It is designed with a set of NO contacts that come into contact with each other when the switch is moved or shaken. The switch can also be of the NC type, with the pendulum breaking the circuit whenever it is moved or shaken.

Although the sensitivity of these older sensors can be manually adjusted by turning a screw, they have a tendency to produce false alarms. The reason is that low-frequency noise, such as that from heavy traffic and many other sources, will create enough vibration to set off the alarm. For this reason, a new type of vibration detector, the piezoelectric sensor, has been introduced. In appearance, it is similar to the older types.

This is a sensor that is sensitive only to high-frequency vibrations, such as those emitted by breaking objects. The sensor contains piezoelectric crystals, a crystalline material that will develop a voltage when subjected to mechanical stress or severe vibration. The voltage will trigger the alarm. This sensor is often used on walls, windows, and doors. The sensitivity can be adjusted, and the chosen sensitivity will be remembered by a memory circuit. As semiconductors are used, the piezoelectric sensor has no moving parts. The piezoelectric sensor is, in fact, a kind of glass breakage detector but of a type that is not limited to being affixed to the window pane.

Similar devices work according to a different principle. These are acoustical detectors mounted near windows, generally in the ceiling. The detector is then able to guard a number of windowpanes within its range. This detector is sensitive to the noise of breaking glass, shattering wood, and other types of noise signifying a break-in. Its range is often around 15 meters. The detector contains an advanced digital filter, which will ensure that lower frequency background noise will not trigger the alarm. The filter will usually set the lower limit at 5 kHz, but this is often adjustable to between 2 to 10 kHz. Likewise, noise that builds up gradually, such as from cars, aircraft, and vehicle brakes, will not trigger the alarm. This prevents false alarms to a large degree.

The system works on the principle of audio discrimination. This is the process of electronically separating normal everyday sounds, such as voices, telephones, etc., from break-in type noises such as are caused by breaking glass, prying metal, or forcing a door open.

Highly sophisticated sound detectors can easily respond to the sound of a window being smashed. The detector will then respond to the brief time lapse between the sound of the window being smashed and the tinkle of falling shattered glass. Some of these detectors are so sophisticated that they can even distinguish between the sounds of a breaking window and a breaking bottle. There might also be a listening-in function that allows a central security station to listen in on what is happening in the room through the use of an automatic dialer or some similar equipment. This is made possible by the electret microphone built into the sensor. These alarm sensors generally—but not always—work with NC switches.

The most recent type of vibration detector is the inertia sensor. This device looks virtually identical to the other types of vibration detectors. It is an intelligent vibration detector and, consequently, requires a special control unit to analyze the complex signals from the sensor. The principle of this sensor is a comparatively heavy contact element, which rests on a contact surface that is mechanically connected to the cover of the detector. The contact element will not vibrate with the frequencies that are emitted when glass is broken or when metal strikes glass. The rest of the sensor will, however.

Higher frequencies therefore give rise to a burst of short-duration breaks, which are analyzed in the control unit. When the number of short-duration breaks reach and exceed a threshold value, the alarm will be triggered. This will only happen when the vibrations are strong enough to indicate a real break-in attempt. Clearly, determining the likelihood of a real intrusion is a very complicated process, and only a comparatively complex control unit can do so. Therefore, these sensors have other control units in addition to the central control unit of the alarm system.

Inertia sensors are also often found in cars, caravans, and on containers. One such sensor, with its special control unit, is sufficient to guard the entire perimeter (i.e., the shell of the construction). When properly installed, the inertia sensor is a very reliable alarm sensor.

INFRASOUND DETECTORS

This is a fairly new type of detector, sometimes used for perimeter alarms in houses and in other relatively enclosed objects, with a floor space not larger than 400 square meters. One detector is sufficient even in a house with several floors, as long as the total area is not too big. It works by detecting sound within the frequency range of 1 to 4 Hz. This is called infrasound, as it is below the sound level normally audible to the human ear. Such sound appears because of the change in air pressure that takes place when a window or a door is being opened. Changes in the air pressure always take place when the enclosing material changes its nature.

It is perfectly possible to move around in the building when the alarm system is armed, as long as no door or window is opened. As only one sensor is necessary, this alarm is very quick and easy to install.

The disadvantage of this system is that it is a one-time alarm only. If an intruder leaves the door or window open, the alarm will only sound once—when the door or window is first opened. Infrasound detectors are therefore usually used with other types of trap alarms.

The infrasound detector can be hidden almost anywhere—under a staircase or behind a cupboard or a curtain, for instance. A good sensor of this type will automatically compensate for natural infrasounds, such as those emanating from strong winds.

FIELD EFFECT SENSORS

The field effect sensor relies on the principle of capacity changes between the guarded object and earth or between extended conductors or foils. In effect, it works like a radio transmitter and receiver.

If somebody is moving in the field between the transmitter and the receiver, there will be interferences in the received signal. This will trigger the alarm. For this reason, these sensors are sometimes known as field disturbance sensors.

Capacity changes can be indicated in different ways. If a metal safe is to be protected, it can be connected to a frequency-determining resonance circuit. This resonance circuit will control an oscillator, whose output frequency will be compared to a fixed frequency within the range of 100 kHz to 10 MHz. The variations in frequency will be detected and analyzed. If a predetermined threshold value is exceeded, the alarm will sound.

Likewise, a building can be fitted with two encircling conductors, approximately 1 meter apart. One will be the transmitting antenna, in effect the antenna to a long wave radio transmitter, while the other will be the receiving antenna (fig. 82). The system will detect and analyze the differences in the received signal caused by an intruder approaching the protected area.

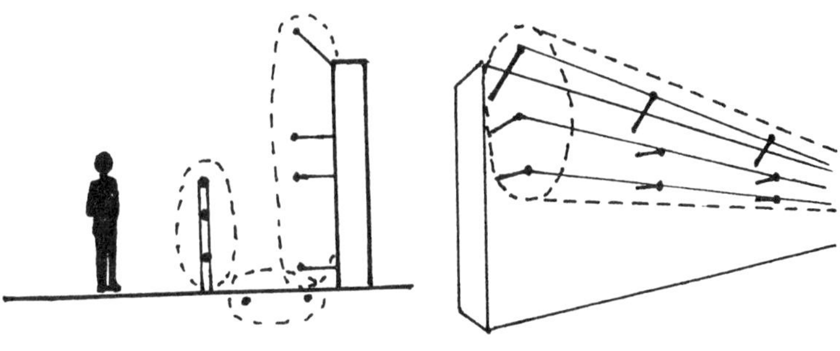

Figure 82. Field effect sensors.

Field effect sensors generally consist of two or more wires running parallel along the protected area. The wires will be connected to the wall or fence by insulators. Particular care will have been taken to position the wires far away from other metal objects, as such objects may limit the range of the detector. Walls, internal walls, corridors, and other locations can be protected in this way. In a corridor, for instance, the two wires up to approximately 300 meters long will run one on each side. Alarm systems of this type will definitely be more common in the future, and they already exist in many places.

Another variant is the capacitance detector. Such a detector often consists of a metal plate on which the protected object is positioned. If an intruder approaches the object and the metal plate too closely, the electrical capacitance in the plate will change and trigger the alarm.

Sound Detectors and Heat Detectors

Freestanding, portable sound detectors can be placed almost anywhere and are extremely easy to activate. Some suspicious individuals put one on a table before they go to bed. The sensor will listen for intruders in the room where it is located, and possibly adjacent rooms, too, as long as the internal doors are left open.

Sound detectors are very prone to false alarms, however. The problem is basically that they are too good. They can be triggered by perfectly ordinary noise outside the protected area, for instance, in the street. The most common type of sound detector is the previously described glass breakage detector.

Sound detectors are often mounted on safes. They sound the alarm if some-

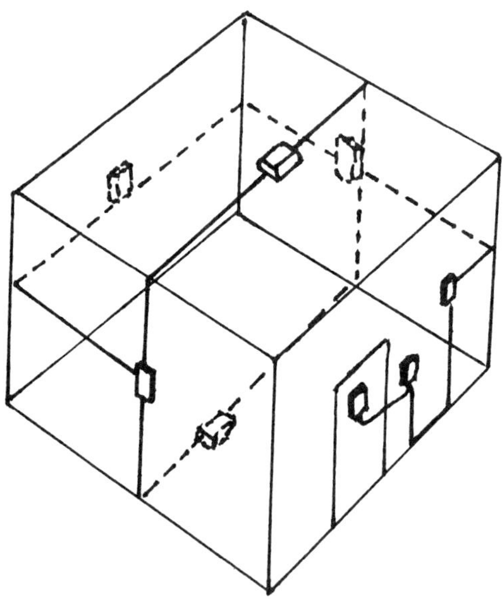

Figure 83. Sound detectors mounted in a bank vault.

body is working on the safe in an attempt to crack it. This detector consists of one or more microphones connected to an amplifier, programmed to only respond to the type of noise expected to signify a break-in. This will ensure that the occasional noise of a broom or vacuum cleaner hitting the safe during cleaning will not trigger the alarm. The microphones are attached to the object with a strong magnet, and a plunger switch or magnetic reed switch sabotage alarm is included, so the alarm will sound only if somebody attempts to remove the detector from the object. Sound detectors are commonly used in bank vaults and are mounted on every wall, as well as in the floor and ceiling (fig. 83).

A sound detector is often mounted on a safe along with a heat detector. This device should not be confused with the infrared detector. The heat detector will only register the heat of a welding torch or a fusing burner.

Pressure Mats, Plunger Switches, and Contact Strips

Pressure mats (fig. 84) look like rubber floor mats and are commonly hidden under wall-to-wall carpet and

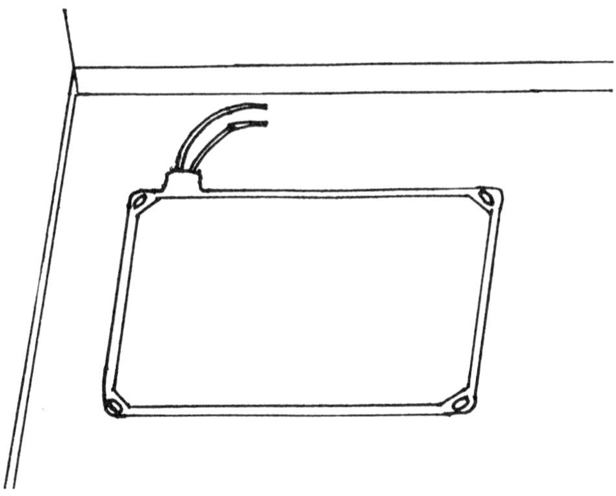

Figure 84. Pressure mat.

linoleum floors at strategic places in the building (e.g., in front of certain doors and windows, at the foot of the stairs, or directly in front of a safe or important object). It is also common to use a series of pressure mats in a staircase, so that it is virtually impossible for an intruder to avoid stepping on at least one of them if he wants to proceed to the next floor. As staircases are often of different sizes, the pressure mat also comes in several different sizes.

The pressure mat is in fact a large NO switch. The mat contains two grids of switch contacts separated by a nonconductive material that has been perforated several times. The contacts can be brought together through the holes. The contacts are only pressed together when a person of sufficient weight (i.e., a person; children will trigger the contacts, but not pets such as cats or small dogs) steps on it. Then the NO switch is closed and the alarm triggered.

If the pressure mat is hidden in an unsatisfactory way, the outline of it might start to show through the carpet after a long period of use. The intruder can then avoid it easily. Certain inferior brands of pressure mats will also be triggered by pets such as cats or dogs, so the presence of these animals might indicate the absence of pressure mats, at least in older or less secure buildings.

A plunger switch is a mechanical device located on doors and windows to detect entry or tampering (fig. 85). It is an ordinary spring-loaded momentary switch that is designed for use in

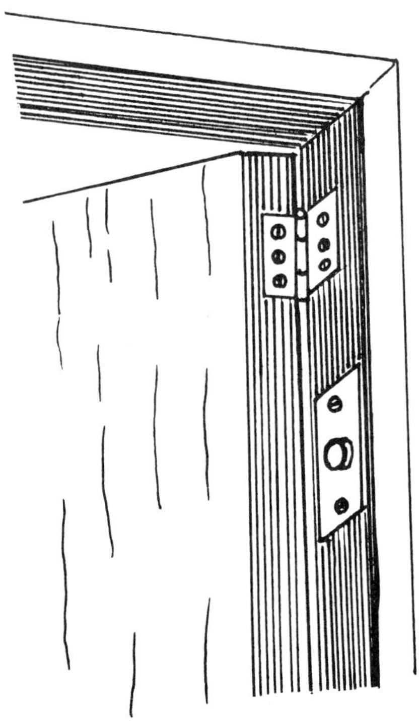

Figure 85. Plunger switch.

alarm systems and has a depressible plunger or button. The plunger in itself is oversized, while the body of the switch is designed for mounting on doors, windows, or control units. The plunger switch is easy to bypass, as long as it is known to exist. Simply keep the plunger under pressure with a piece of celluloid, for instance, so that the device does not indicate an alarm while you are removing the switch from the protected area.

The plunger switch is often used in control units to prevent the system from being tampered with. In this application, switches will be mounted on the front and/or back of the control unit. The alarm will sound if the door is opened or the control unit is removed from the wall. Here it is more difficult to remove the switch because of its location.

In the same way, plunger switches can be used to detect when a door or window is opened. In this case, the switch is mounted so that the plunger will be depressed when the door or window is closed. The switch is then mounted on the hinge side in the door jamb and is virtually impossible to see, as it is neither exposed from the inside or the outside when the door is closed. Of course, this type of switch is an NC switch. Whenever the door is opened, the alarm will sound.

A contact strip mechanism is, in effect, an open switch and is sometimes used on windows and doors, usually together with window foil and glass breakage detectors.

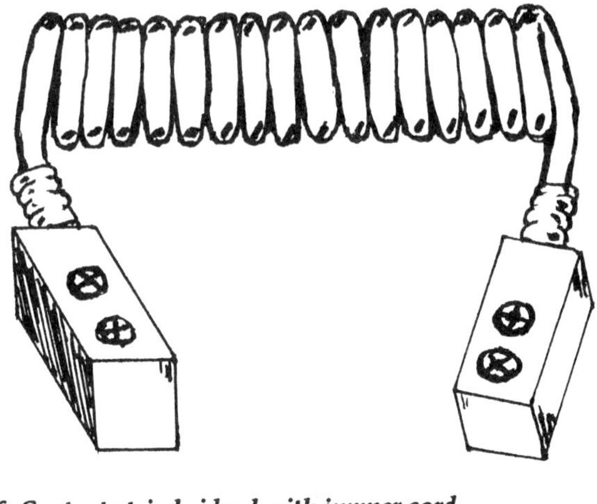

Figure 86. Contact strip bridged with jumper cord.

The contact strip is similar in shape to the reed switch and is also composed of two parts. Here, too, one part is mounted on the window or door, while the other is mounted on the frame (fig. 86).

The advantage of the contact strip is that the two pieces can be bridged together with a jumper cord. This is highly useful when a window or door (e.g., a garage door) must be left open but the alarm system is still armed.

Ionization Detectors

The ionization detector is not used anywhere yet, as far as it is known, but this might well be the alarm sensor of the future.

The ionization detector works according to the principle of the Kirlian effect, which states that living matter is surrounded by a "force field" that ionizes the surrounding atmosphere. This will produce an aura, or halo, around a living being photographed in fields of electrical current. (The Kirlian effect is not universally accepted, as the reason for this phenomenon is so far unknown. The phenomenon has also been abused frequently by sensation-seeking parapsychologists. However, this has not prevented research in ionization detectors.)

The ionization detector will work in the following way. It is well known that if the atmosphere is ionized this will change the electrical conductivity. An ionization detector will therefore register the atmospheric conductivity. Any changes resulting from the ionization of the surrounding atmosphere that is caused by the "force field" around a human being will be detected.

No known alarm systems rely on this principle today, but in what was formerly the Soviet Union, the Kirlian effect has been accepted for several years. It is probably only a matter of time before sensors of this type are in production.

Photoelectric Cells and Invisible Beam Detectors

Photoelectric cells have been in common use for years, especially in shops, to detect a person walking into a room or through a door. Old-fashioned systems of this type use visible light, and the person entering has no trouble at all noticing and evading the photoelectric cells should he choose to do so. Nowadays, what is basically the same system is still widely used, but instead of visible light, invisible infrared light is used.

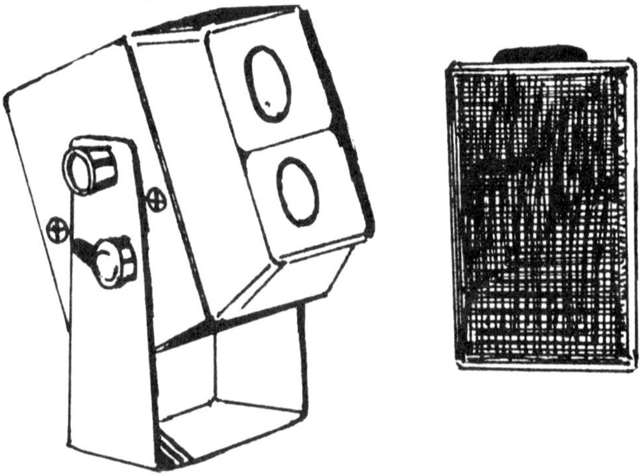

Figure 87. Photo relay sensor with detector (left) and the reflector (right).

The modern invisible beam detector, or photo relay sensor (fig. 87), is designed to project a narrow infrared beam across the area to be protected. For this reason, sensors of this type are also sometimes called active IR-detectors. The beam is aimed onto a small reflector, which will reflect the infrared beam back to the invisible beam detector, which has a built-in photoelectronic eye that is sensitive to infrared light. This photoelectronic eye will constantly monitor the area, and as soon as somebody interrupts the beam between its projector and receiver, either completely or almost completely, the alarm sounds.

Most invisible beam detectors of this kind are able to protect an open area up to about 10 meters wide. Larger units, however, are conceivably capable of a range of several hundred meters. These units are commonly used outside a building, in a private garden or park, for instance (fig. 88). Sometimes they are used to protect rooftops as well. With

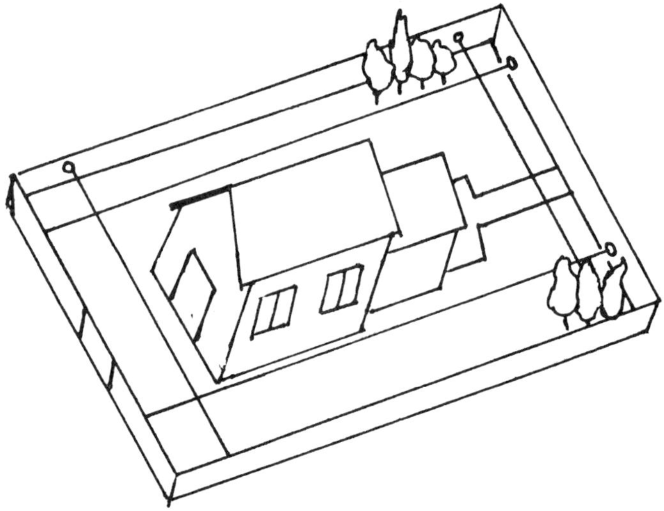

Figure 88. Infrared beams arranged for perimeter protection.

the exception of really sensitive installations, these detectors will generally trigger security lights rather than sirens, as the chance of false alarm is very high.

The most common method of installation is with two invisible parallel beams. Then a small animal such as a bird will not trigger the alarm, but a human-sized intruder will break both beams and trigger it. Likewise, mist or falling snow, rain, or leaves will not trigger the alarm. The range is generally up to 150 meters, but it is shorter in countries with cold climates. Many detectors of this type can function even if covered by snow or frost, however.

The emitted infrared light is modulated so that the receiver can identify it without being disturbed by other

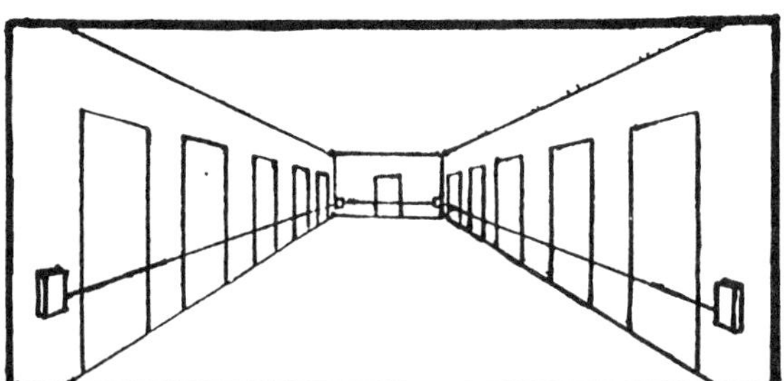

Figure 89. Infrared reflectors used to protect a corridor.

sources of infrared light, such as sunlight. The infrared beam might also be reflected by infrared reflectors (fig. 89), which will increase the range even more. As the invisible light used in the beam consists of infrared light, these detectors can be found and evaded easily by using infrared-goggles during the entry operation.

Passive Infrared Detectors

A passive infrared detector, or PIR as it is commonly known in Britain and some other countries, is a device that receives and measures infrared energy from other objects. It is, in effect, a heat detector, although it is of a very different type than the heat detectors mentioned earlier in this chapter.

The passive infrared detector is usually fixed high on the wall, either in a hall or in a large room which must be traversed to reach the sensitive areas. It works by picking up an intruder's body heat, triggering the alarm whenever it detects body heat within its range. The detector is extremely sensitive and will pick up even the smallest variation in temperature, both above and below normal.

The passive infrared detector works by utilizing a built-in pyroelectric sensor. Simply speaking, this sensor is able to indicate heat by producing an electric charge. The pyroelectric sensing material is polarized by infrared radiation producing a voltage proportional to the rate of change of incident energy. The pyroelectric sensor will electronically monitor the room protected by the alarm. It is surrounded by a reflective surface that collects infrared energy. As the sensor is sensitive to heat (i.e., infrared energy), it will sense any change in the level of infrared energy, including the presence of any intruder radiating heat. Such an abrupt change, such as is caused by the appearance of an intruder, will be readily detected and will trigger the alarm.

A resting human will radiate approximately 100 W, so we are all major sources of infrared radiation. However, mere radiation is not enough to trigger a passive infrared detector. The heat source must also be moving through the zone monitored by the detector.

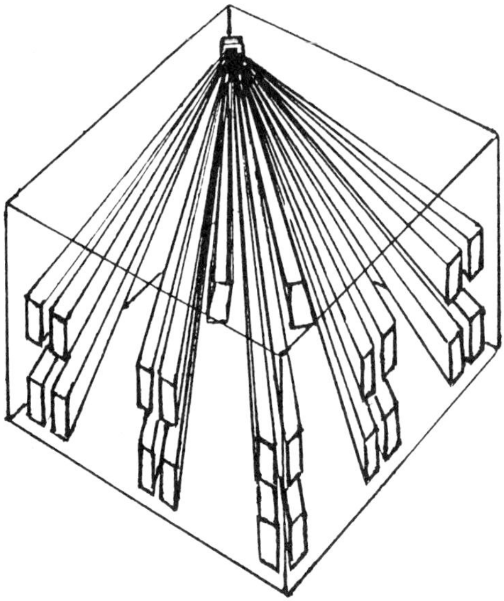

Figure 90. The detection zones of a passive infrared detector.

The passive infrared detector will divide the area to be monitored into zones and segments (fig. 90), typically up to 7 meters across (actually 76 degrees wide, usually, but this can vary), 12 meters deep, and around 70 centimeters high. Only these zones and segments will be monitored, and intruders above or below them will not be detected. The detector must therefore be adjusted very carefully during installation to ensure maximum protection as well as to prevent false alarms triggered by pets or small children. For these reasons, it will generally be positioned at the level of a man-sized intruder's head or higher—at least 1 1/2 meters above the floor. The detector is most often positioned along a wall or in a corner so that it can have a free line of sight to as much of the room as possible, including the front door, if at all feasible.

A passive infrared detector typically looks at a zone of six or more separate segments with unwatched "aisles" between them. A separate-zone design gives the unit wide-angle coverage of a room. As the background radiation of each zone is slightly different, it is virtually impossible for an intruder to match them all.

Self-contained infrared sensors generally cover a wider area and have more zones but typically protect an area up to only 8 meters deep. There is also a lower zone in each segment. This will usually ensure that an intruder cannot slip under the monitored zone. Sometimes, especially if pets are on the premises, the lower zone will be turned off.

There are also infrared detectors on the market that can monitor areas up to 50 meters in length. The range of some of these can be varied if the owner changes the lenses.

The IR-detector is prone to false alarms, however, as it will be triggered not only by obvious problem sources such as cats and dogs, but also by small pests or bright sunlight through a nearby window. Even being too near any quick-changing source of heat or cold, such as a central heating radiator producing radiated heat and warm draughts or an air-conditioning duct, might produce a false alarm.

When an infrared detector is switched on, it first balances itself based on the amount of background infrared radiation coming from various sources in a room, such walls, furniture, and floors. If an intruder later enters the detection zone, he alters the amount of infrared radiation detected. This results in an alarm.

This need to first balance the sensor means that if the power has been switched off for some time, the sensor will require approximately ninety seconds to be operational. However, this is not necessary when the system has merely been disarmed but has retained its power.

In addition, the detector will not respond to slow variations in background radiation because amplifying circuits limit the detectable variations to a predetermined range of possible speeds.

In advanced units of this type, a threshold circuit will also ensure that a signal is large enough to represent an intruder. In this case, a pet will not trigger the alarm, as long as it is not very close to the detector. Of course, the sensor's sensitivity to false alarms is also determined by the positioning of the detector and its segments of protection.

Every infrared sensor will be most effective in detecting movement across the segments, rather than movement toward or away from them (fig. 91). There have been lab tests to determine the possibility of making a very slow direct approach to the sensor so that it will not feel a sharp increase in temperature and, consequently, refrain from sounding the

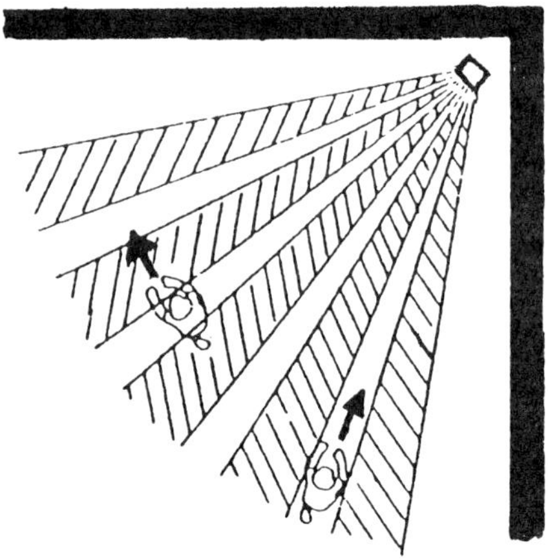

Figure 91. The passive infrared detector is most sensitive to movement across its segments.

alarm. In some of these tests, the "intruder" was hiding behind a thick glass pane, as plain glass will absorb a considerable portion of the radiated heat. However, no such experiments have been deemed successful enough to warrant the use of the techniques involved in the field.

An infrared detector can also be mounted in the ceiling. If mounted as high as 6 meters, the protected area will be as large as 6 x 20 meters, and all movement will naturally occur across the segments. This facilitates detection.

Passive infrared detectors, although generally used to monitor movement in rooms or along corridors, can be used vertically to create an alarm curtain, in front of a wall of paintings, for instance. They can naturally also be used to create a horizontal curtain to detect a break-in through the roof, ceiling, or floor.

Most passive infrared detectors have built-in tamper switches to prevent somebody from removing the unit from the wall or attempting to disarm it. Infrared sensor systems, whether self-contained or part of a separate-components system, are easy to locate. The sensor is generally positioned so that its zones of detection include the expected entry, such as doors and windows. The intruder should be forced by the layout of the building to walk into these zones. Furthermore, the sensor will be placed at least 1 1/2 meters above the floor. This is necessary if the lower portions of the zone are to be used.

The sensor unit will be positioned so that its field of view does not include solar-heated walls, direct sunlight, heaters, air conditioners, or other objects that might change its temperature quickly. The older infrared sensors might produce false alarms under these circumstances. Contemporary models will not do so as readily, but their detection ability will be impaired somewhat. In the contemporary models, every segment is divided into two channels. The detector will only sound the alarm if the change in temperature is different in both channels. This will prevent the alarm from sounding when a radiator heats up, for instance. An intruder, however, will first disturb one channel and then the next, thus triggering the alarm.

Self-contained infrared sensor systems incorporate entry and exit delays, typically fifteen to twenty seconds in length. Otherwise, a remote key switch might be used. Backup batteries can also generally be connected, as can external warning devices.

There are also smaller, self-contained infrared detectors. They generally work with only one coverage zone and are supposed to guard only small areas near doors, windows, trailers, etc. If a heat source is detected, the built-in siren will sound for about one minute, after which the unit will reset. An entrance and exit delay of a few seconds is generally built into these units. Power is generally provided by a standard 9-volt battery located inside the detector unit.

Microwave Motion Detectors

A microwave motion detector is a trap alarm with a single-unit transmitter/receiver that reacts to distortions in the timing of its return signal. The device works, in effect, like a small radar system, using an electromagnetic field comprised of high-frequency radio waves generated over the protected area. Most detectors of this type have a range of 30 meters or less.

The microwave motion detector relies upon microwaves, very high-frequency radio signals in the range of around 9 GHz. These radio waves will form a pear-shaped lobe (fig. 92) that will detect any intruder. As these microwaves are reflected off solid objects, they will reveal the presence of an

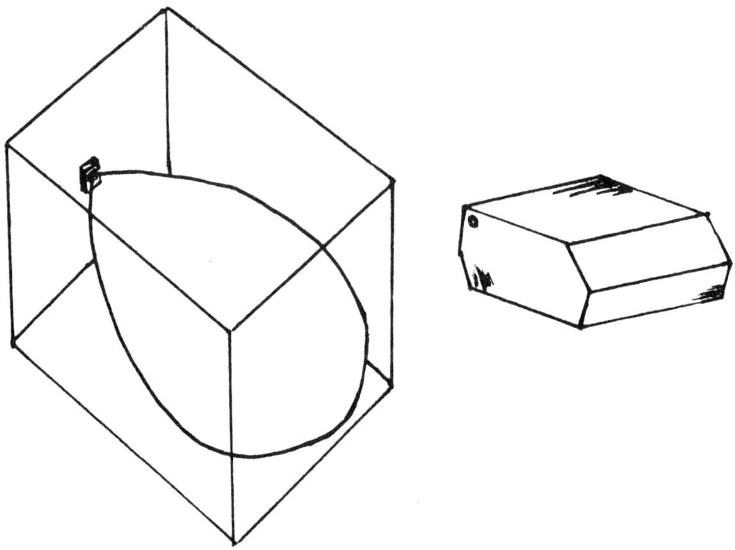

Figure 92. Microwave motion detector and its detection zone.

intruder entering the protected area, as his movements will cause a disturbance in the reflected radio waves (i.e., the radiated RF field sensed by the device).

This disturbance is a modulation of the field referred to as the Doppler effect. The Doppler effect is the difference in frequency of one (original) wave to another (reflected) wave that is superimposed upon it. The reason for this is that a frequency shift occurs when a signal source and a receiver are moved relative to each other. According to the Doppler effect, the reflected signal will be of a lower frequency than the emitted signal if a reflecting object or a human is moving away from the detector. Likewise, the reflected signal will be of a higher frequency than the emitted signal if a human is approaching the detector. The reason for this phenomenon is that since the signal moves at a constant rate, the return trip of the reflected signal should take the same length of time as the outward trip. However, an intruder who moves into the path of the signal distorts the timing of the return signal. This frequency shift is sensed by the microwave detector unit's receiver. Whenever it detects

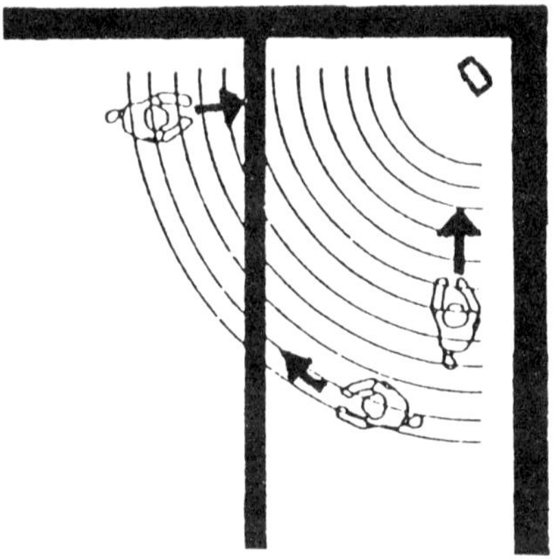

Figure 93. The microwave motion detector is most sensitive to movement toward or away from it. Note that it is also sensitive to movement in another room.

such a change in the reflected signal, the unit will trigger the alarm.

As the frequency of the microwaves is very high, the wavelength is very short. This means that even a very small movement will result in a large frequency shift. However, the detector is less sensitive to objects moving across its field of detection than it is to objects moving toward or away from it (fig. 93).

Most microwave detectors require that an intruder move at a certain speed in order to be identified as an intruder. These detectors are able to reject faster-than-walking speed. This will prevent numerous false alarms, as many microwave motion detectors, especially the older models, work all too well. As their signals can penetrate beyond the walls of a house, they might identify a passing truck as an intruder, for instance.

External radio sources can also trip these units, as the RF energy radiated by many electrical devices will disturb their functioning. Such electrical devices include radios, especially citizen band radios, but also cable television systems, motors, transformers, and even neon signs and fluorescent lamps. The gas in such lamps, when switched on, ionizes to become a fluctuating reflector which can easily cause an alarm. A false alarm might also be caused by cooling fans and sometimes even the roof in a warehouse, if the roof rises and falls in the wind. The movement of water in pipes might even trigger a false alarm, as the sensor has the ability to see through glass and thin walls of plaster or plastic.

As the microwave motion detector is an extremely small

radar, the microwaves are completely harmless, unlike the radiated energy from a large radar such as is used for air traffic control or military purposes. There are also microwave/passive infrared detectors. They are very reliable, as the two sensor types complement each other well. The microwave detector determines that the object moves within a certain range of speed, while the infrared detector checks whether the object emits heat or not. These combination detectors come in different types, with ranges up to 65 meters.

Ultrasonic Motion Detectors

An ultrasonic motion detector is a single-unit transmitter/receiver that signals an alarm when its steady pattern of inaudible sound waves is disturbed. It works in a similar fashion to the microwave sensor, but it uses ultrasonic sound waves instead of microwaves. The principle is still the same as that of radar.

The ultrasonic detector works by continuously transmitting ultrasonic sound waves of such a high frequency, around 40 kHz, that a human cannot hear them. The typical upper limit of human hearing is twenty thousand cycles per second, or 20 kHz. The ultrasonic sound waves are therefore well beyond the human range of hearing.

The sound waves will bounce off the hard surfaces in the room, in effect producing an echo which, thus reflected, will be picked up again by the same unit. An intruder entering the room will interfere with the frequency of the reflected sound. Because of the Doppler effect, the reflected sound waves will experience a frequency shift if they are reflected from an intruder entering the protected area. When the intruder moves within the field, the unit's receiver detects a change in the reflected signal and triggers the alarm.

Because of the nature of sound, the ultrasonic detector will produce a teardrop-shaped "cone" of ultrasonic energy, similar to what is illustrated in Figure 92. This cone will expand horizontally and vertically away from the unit across a relatively broad area, typically up to 9 or 10 meters in a forward direction and up to 7 meters wide. However, the shape and size of the cone will vary with the acoustical characteristics and shape of the room in which the sensor is located. Walls and glass windows will not be penetrated, except to a minor extent. There are also ultrasonic detectors that work with separate transmitters and receivers. They often have a longer range.

The ultrasonic detector is most sensitive to movements directly toward or away from the sensor (fig. 94). This is unlike the passive infrared detector, which is most sensitive to movements across its protected area, but similar to the microwave motion detector. The reason, of course, is that both of these types of detector rely on the Doppler effect. Ultrasonic detectors are therefore used more commonly for protecting doors and hallways than are passive infrared detectors.

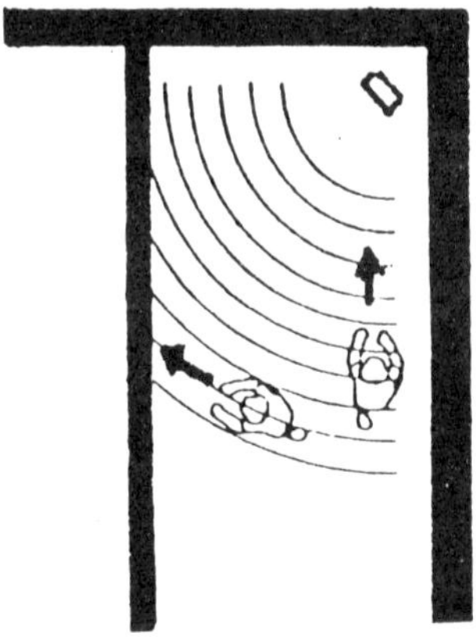

Figure 94. The ultrasonic motion detector is also most sensitive to movement directly toward or away from the sensor.

Ultrasonic detectors are usually located in the corner of the room to be protected, fixed to the wall, or else aimed down a hallway or at an entrance, such as the front door.

These detectors are fairly prone to false alarms, which might be produced by loud noises, such as ringing telephones, as well as moving objects, such as draperies near vents or fans. Even air currents, draughts of moving air from heating or cooling vents, and other types of air turbulence can produce false alarms. For these reasons, most ultrasonic detectors include a control for adjusting the sensitivity of the device. The ultrasonic detector cannot be used if pets are present, though, as there are no unprotected areas within the coverage zone.

Many ultrasonic detectors have a delayed alarm, which

makes the intruder's first steps into the protected area appear to go undetected until it is already too late.

Because of their tendency to produce false alarms, ultrasonic detectors seem to be decreasing in popularity.

VISIBLE LIGHT DETECTORS

Visible light detectors are not common, and they can only be used in certain locations. They are mainly used in conjunction with other detectors in bank vaults. The detector will simply sense if the level of visible light in the location is rising. If so, it will sound the alarm. (The bank vault is, of course, supposed to stay dark when closed.) A detector of this kind is of a very simple design. The visible light detector can only be used in closed locations where no natural light can enter.

VIDEO DETECTORS

A video detector constantly monitors the object it is guarding by merely "watching" it. The detector, a modified video camera, checks the level of black (the number of black dots, opposed to the number of white dots, in a picture) in certain interesting areas of the video input. If this level is changing slowly, the detector will not trigger the alarm, as this might indicate sunset, for instance. However, a drastic and immediate change will trigger the alarm, as this means that a person or an object has entered the field of vision and is now near the protected object.

Video detectors are reliable but fairly expensive. For this reason, they are not widely used as yet. They will probably be more common in the future, particularly since it is also possible to transmit the video signal across the telephone network by means of a modem. Then a remote control station can view and interpret the video.

The latest detectors of this type are even more advanced than the standard video detector. A number of cameras can be used to constantly monitor the zone to be protected. The video output from each camera will then be processed by an intelligent image processor based on neural network technology. This will automatically identify any intruder.

The system works by first learning the characteristics of the natural state of the environment under observation, including moving items such as shadows, branches of trees, or level crossing barriers. This initial learning state does not need to last for more than a minute. The system can even be taught to ignore certain dynamic events, such as guard patrols or vehicular movement.

When the system enters operational mode, it identifies intruders within its field of view by recognizing abnormal patterns of movement. It immediately sounds the alarm and automatically trains a high-resolution camera onto the target for identification and video recording purposes.

This advanced system is not yet in widespread use, but it can be found in certain British defense research installations. It can also be configured to be portable. Expect it to be common in most high-risk installations in the future.

Barrier Sensors and Analyzers

A perimeter barrier is a wall, fence, or gate marking the perimeter of the property. Guarding such a perimeter, especially if large distances are involved, requires special sensors and sometimes also special control units or analyzers.

Today, the most popular barrier sensor is the inertia sensor, even though there are various other types on the market. The inertia sensor, used in conjunction with a special barrier analyzer, can be adapted easily for use on any fence.

The inertia barrier sensor system is an electromechanical system that relies on special wires (fig. 95). Every wire is connected to a special self-adjusting sensor installed in a sensor post, or pole. The sensor is positioned between two horizontal wires that run between the sensor posts. The sensor will notice immediately any attempt to spread the wires (to enter between them), climb on them, cut them, or otherwise remove them.

The sensor-connected wires are fitted with vertical springs and self-adjusting connection points so that they

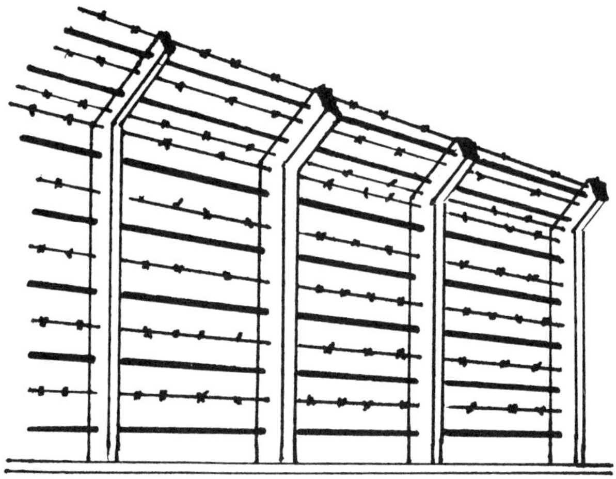

Figure 95. Inertia barrier sensor used on a fence.

will detect any kind of intrusion. Six wires and sensors will generally be fixed parallel to each other on the posts so that it is impossible for an intruder to avoid all of them if he is attempting to enter. This also allows the system to determine the exact height of an intrusion. Another advantage is that a wire group can be disconnected, during maintenance, for instance, without the rest of the sensor post being affected or impaired.

The sensor posts are generally positioned up to 50 meters apart, although a shorter distance, say 10 meters or so, will make the system able to really pinpoint an attempted intrusion. The sensor posts are made of aluminum or stainless steel, unlike the ordinary posts that are made of stainless or galvanized steel. Every sensor post has individual tamper protection and is also monitored individually by the computer in the central security station.

A multiplex computer communications system will keep in touch with every single sensor post and wire through a special information cable that runs along the fence. Every detected intrusion will then be reported by a special reporting unit in each sensor post and will be monitored and registered by a computer in the central security station. The system will also self-test continuously, sounding an internal alarm if any part of it breaks down or is subjected to attempted sabotage. Naturally, this self-test also includes the information cable. In addition, every incident will generally be logged in the central security station computer if the need arises later to verify what happened.

The computer communication will go in both directions, which means that individual security measures can be adopted in case of an alarm. These might include video cameras and searchlights or armed mines programmed to be activated automatically, without the need for a manual operator.

As can be readily imagined, this type of system is very complex and only used in high-risk locations. The most famous system of this type is the Israeli Magal system. Magal Security Systems now guard the entire length of Israeli border fences and barriers. An identical system, also produced by Magal, is used in numerous other countries, in military installations and around nuclear power plants, for instance. The best way to circumvent this alarm system is to avoid touching the fence. Try to enter in another location or from the air.

Another common sensor used to protect a perimeter is

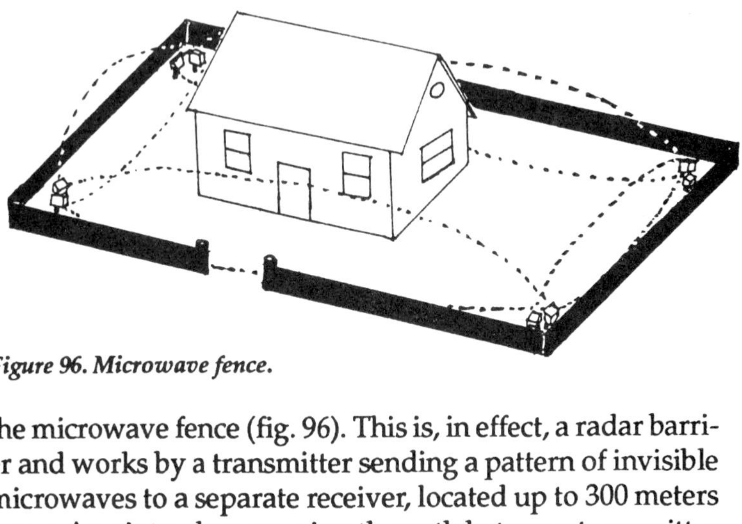

Figure 96. Microwave fence.

the microwave fence (fig. 96). This is, in effect, a radar barrier and works by a transmitter sending a pattern of invisible microwaves to a separate receiver, located up to 300 meters away. Any intruders crossing the path between transmitter and receiver will distort this pattern and the alarm will then be signaled.

Transmitters and receivers can be positioned along the perimeter of a garden, for instance, or any other open area. The microwaves will create a signal field, which might be between 2 and 8 meters high and up to 20 meters wide or more.

This sensor is very prone to false alarms, as there is no way of distinguishing between an intruder and an ordinary bird or animal. Even moving vegetation might trigger it. It will, however, detect even small changes, so an intruder crawling slowly into the field will be detected. The sensitivity is adjustable, however, and some systems of this type will only indicate an intruder that is at least walking slowly. The speed of movement that the sensor can detect can usually be adjusted to between .01 and 10 meters per second.

As the transmitter and receiver will be connected by a synchronizer cable in order to force the receiver to notice only its dedicated transmitter, the system can be sabotaged by cutting this cable. However, this will generally also trigger the sabotage alarm. The system cannot be sabotaged by simply transmitting microwaves of the correct frequency into the receiver, as the synchronizer will reveal this as an error.

Microwave barriers are generally fenced in, as the presence of wild animals would otherwise give rise to frequent false alarms.

Yet another device used for external alarms is the geophone (fig. 97). This is a device that monitors vibrations. It can be installed to detect activity across the ground or the vibration caused by the scaling or attacking of walls and fences. The sensor is, in effect, a vibration sensor.

An older and cheaper barrier alarm system relies on mercury switches as fence alarms. The switches are mounted on the fence and will sound the alarm whenever the fence is moved by somebody climbing it or leaning a ladder against it. The mercury switch is also a vibration sensor, although of an older type. A mercury switch fence can also be circumvented by not touching the fence.

Field effect sensors can also be used as barrier alarm sensors. The system will then use wires connected to the

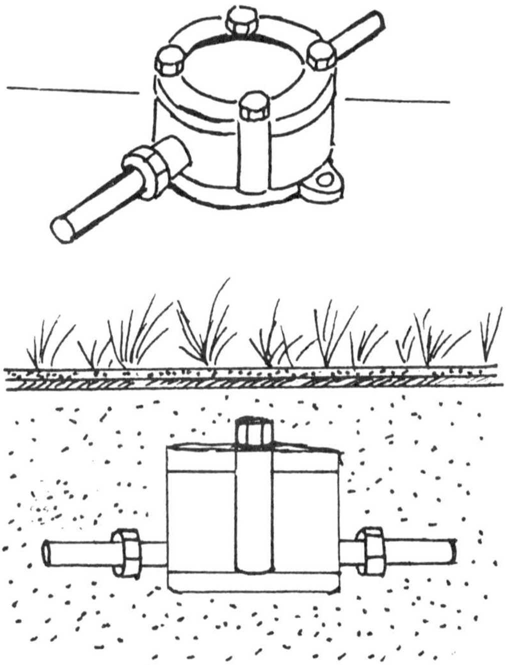

Figure 97. Geophone.

top of the fence. Otherwise, it is the same as the field effect sensor detailed earlier in this chapter.

Yet another system is the cable detector. Cable detectors consist of special cable that is capable of changing resistance or capacitance when bent or hit. The cable will run through the important parts of the fence. It will trigger the alarm if it is bent, climbed on, or cut by an intruder in order to facilitate an entry. Any such movement of the fence will cause a capacitive change in the cable that will in turn trigger the alarm.

Another fairly reliable system uses a thin tube or hose connected to a microphone. The signals from the microphone will be analyzed by an advanced electronic circuit. Any attempt to climb or destroy the fence will be recognized and trigger an alarm. The microphone can be connected to a loudspeaker and monitored in an alarm control station.

All sensors of this type frequently produce false alarms.

Both animals and strong winds that cause the fence to move might be causes. These are therefore usually connected to the control unit separately from other types of sensors, so that the entire alarm system is not tripped just because of an indication from the barrier sensor.

CHAPTER 8

Other Types of Alarm Systems

Deliberately Activated Alarm Systems and Personal Attack Alarms

A deliberately activated alarm, or a panic button, as it is often called, is a device designed to set off the alarm when the individual in charge of the alarm feels threatened or is under imminent attack. Panic buttons are always fixed in place, in a bedroom, for instance, or at the teller's position in a bank.

A personal attack alarm is used for a similar purpose, but this device is not fixed to a certain location. Rather, it is carried on the person to be protected by it and is activated by him or her wherever he might be on the premises. Personal attack alarms are, for instance, often carried by medical personnel who regularly have to deal with potentially dangerous patients. They are also sometimes carried by elderly people who fear that they will be attacked by burglars.

Most professional alarm systems, whether for use by private individuals or by corporate or government offices, include one or more panic buttons or personal attack alarms. Alarm systems in private homes might have two panic buttons, one by the front door and another by the bed. Office alarm systems might have any number of these protective devices.

Regardless of the variety of these alarms available, they all work in the same way. The panic buttons are wired to a circuit on the alarm system that is always live, whether or not the alarm has been switched on at the control unit. The user need only press the button at the first indication that an intruder is trying to enter the premises. The button is designed to be easy to find and depress, even in the dark.

The panic buttons used in banks and similar locations are more complicated. The reason is that most police departments will take several actions as soon as they receive the alarm, such as closing down all public transportation systems in the vicinity. Such precautions are very costly and unpopular, of course, and the police departments do not, for obvious reasons, want to resort to them unnecessarily. The panic buttons are therefore designed to be almost impossible to trigger accidentally. They may be either hand- or foot-operated. Usually they must either be activated by using two fingers or by pressing the foot upwards, to minimize the risk of setting them off by accident.

There will also be a discreet indicator installed near the panic button, informing the person activating the alarm that it has indeed been activated. This is usually an LED or a similar light that will remain lit until the alarm is disarmed manually. There must be no question of whether the alarm has been activated or not. There will also generally be a corresponding indicator coupled with a buzzer somewhere in the rear of the office, well out of sight and hearing of the front office where the alarm was set off. This indicator will inform a security officer of the fact that the alarm has been activated. At the same time that the buzzer sounds, the alarm will go to the local police department through an automated dialing system. In addition to this, it is also common for the security officer to verify the alarm by personally telephoning the police.

A personal attack alarm is essentially a wireless panic button. It is designed to be carried on one's person, and, when activated, will trigger the existing alarm system and warning device in the building. Personal attack alarms either rely upon ultrasonic sounds or radio transmissions.

The radio transmitter used as a portable panic button is a fairly obvious design, in effect simply a transmitter capable of activating the alarm system. The transmitter will send a digital code that activates the control unit. The code will tell who activated the alarm but not the location of the trouble. This is a disadvantage, of course. However, the advantage of a radio system of this type is that the range is generally fairly wide. This personal attack alarm can therefore sometimes be relied upon even outside the house.

An ultrasonic personal attack alarm works in a slightly different way. The ultrasonic signal will be received by one of a number of special ultrasonic receivers, one of which is mounted in every room from which an alarm might need to be sent. The ultrasonic signal will activate the receiver, which will in turn trigger the alarm. An LED on the control unit will indicate which receiver triggered the alarm and, consequently, in which room the person who activated it currently is. The identity of the user will not be known, however, if several of these personal attack alarms are in use. Furthermore, the range of this system is much more limited than that of a radio transmitter.

Both these alarm systems can be connected to a personal paging system. The alarm can then be silent, but, for obvious reasons, this is seldom, if ever, the case. If a silent alarm really is desired, then anybody equipped with a personal paging system can be chosen to receive the alarm.

Personal attack alarms should not be confused with the so-called personal alarms sold in many stores. These are small devices designed to emit a painfully loud, high-pitched screeching noise that will surprise an attacker as well as call for help. These alarms are not connected to any alarm system but are sometimes used by people who fear attacks, especially when going out. The efficiency of such an alarm on a deserted street is not very high, however, although it might be sufficient to scare off a mugger.

There are various kinds of personal alarms. Some of them, generally the smallest ones, are activated by compressed air and look like aerosols. Others are powered by batteries or rechargeable power units. Some of them are activated by a button or a trigger, but they only sound as long as the button is pressed or the trigger is squeezed.

This is especially true of the gas-powered types. More reliable personal alarms work on the principle of a hand grenade. The alarm will continue to sound until it is switched off or the power runs out, even if the alarm is dropped to the ground. This is especially helpful if the situation develops into a fistfight.

Some individuals, especially women afraid of rapists, occasionally wear a whistle instead of any of these types of personal alarms. The idea is the same.

Car Alarm Systems

There is a large number of different car alarm systems available on the market today. Some will sound if somebody tries to jimmy open a door or the trunk. Others will be triggered by an attempt to move the car, by towing it away, for instance. Certain alarm systems will even include a remote pager that will signal the owner if his car is being tampered with.

As in ordinary alarm systems, the car alarm includes a control unit, one or more sensors, and a warning device. In a car alarm, however, some of these devices, notably the control unit, will be simpler in design than in the units available for protection of homes and office buildings.

In many cases, the car alarm is at least partially self-contained. No self-contained alarm system is able to protect the entire car, however, including its trunk and hood.

The control unit is often mounted near or on the dashboard, either by using a mounting bracket similar to those used for radios or stereo systems or by mounting it directly to the surface. As the owner generally does not want to give a potential intruder the opportunity to see the control unit, it is often hidden under the dashboard instead, or under a seat, along the firewall, or in the glove compartment. A backup battery is often provided, too, to increase the reliability of the system. Most types of car alarms are operated by either a concealed switch or a key. The sensors used in car alarm sys-

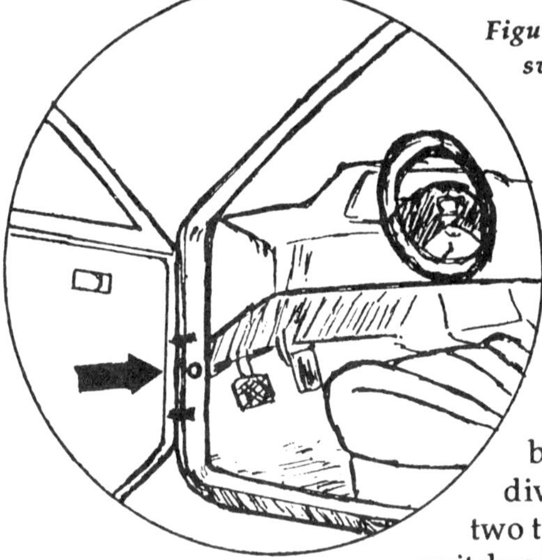

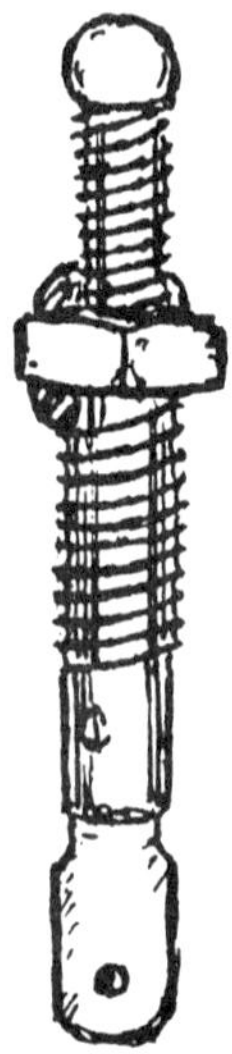

Figure 98. Pin-switch and its location in a car.

tems can be roughly divided into two types: pin-switches that guard the entry points of the vehicle and motion detectors designed to detect movement of the vehicle.

The most common type of car alarm sensor is the pin-switch (fig. 98), also known as the earth-seeking sensor. This is a door contact switch that is activated when the electrical current is broken. A pin-switch is wired directly to the door and installed in the framing around the doors and/or the trunk and the engine compartment. Pin-switches can also be used to guard the trunk, hood, sunroof, tailgate, or any other point of entry.

Pin-switches are spring-loaded, momentary contact switches similar to the plunger switches described in the previous chapter. The pin is released from the switch when the door is opened, which electrically grounds the system, thus triggering the alarm. This device is linked either to the car horn or to an independent siren. An independent siren is a better option, of course, as it is more difficult to find and disconnect.

As pin-switches have only a single wire connected to them (the car's metal chassis being used as the ground portion of the circuit), they are always installed on a metal surface. When used to protect a door, the pin-switch will be installed near the switch for the interior light on the lower part of the door post.

Some car alarm systems use the switches already installed in the door frames of the car instead of pin-switches. These already installed switches are used to turn on the interior lights whenever the door is opened. Electrically, these switches are identical to pin-switches. In this case, the door frame switches are added to the alarm system by attaching wires to them from the control unit.

The other type of car alarm, very commonly used but not really very reliable, is the motion detector. This is a special vibration switch designed to sound the alarm when the car is shaken or moved. This sensor has a tendency to produce false alarms. Nevertheless, it is very popular. Some control units even incorporate such a sensor in their housing in order to protect the control unit itself as well as the car. Such a self-contained car alarm is difficult to remove without triggering the warning device.

If a motion detector is integrated into the control unit, the unit must always be as level as possible, preferably near the center of the car. If positioned in some other location, it will not be as sensitive to motion at the front or back of the vehicle. The most common location is, once again, under the dashboard.

There are many types of motion detectors. One such type is the pendulum alarm. Such an alarm consists of a pendulum switch and is set off when the car is rocked, jolted, or otherwise moved.

A pendulum switch is used to sense vibration or motion. The switch is designed with a set of contacts that touch each other when it is moved or shaken.

The pendulum itself is a small weight on the end of a light spring with a contact underneath the weight. Whenever something causes the spring to vibrate, the weight will touch the contact and the alarm will sound. The sensitivity of the spring can of, course, be adjusted, but most car owners do not do this. The result is that the pendulum alarm is usually triggered by almost anything, such as the vibration of a passing truck or somebody who happens to bump into the car accidentally. Even a very light jolt is sufficient to trigger this alarm. This propensity for false alarms has made the pendulum alarm almost totally useless. In neighborhoods where this type of alarm is in widespread use, nobody ever thinks twice about hearing it and there will definitely be no response to it.

A slightly more advanced version of the pendulum alarm is the trembler switch alarm. This device is similar to the pendulum alarm but instead has a ball bearing sitting between two contacts. Any definite movement of the car, such as somebody trying to open a door, will cause the ball bearing to touch the contacts and trigger the alarm.

Both the pendulum and the trembler switch alarms are generally mounted under the hood on the firewall in the engine compartment and can be adjusted for sensitivity. Another common place to find these motion detectors is under the dashboard.

The vibrator contact circuit, or adjustable impact device, is another motion detector. It has a relatively light weight on the end of a piece of spring steel, with a contact under the weight. When the spring steel vibrates, the weight touches the contact, thus activating the alarm. In this device, sensitivity can be controlled by adjusting the distance between the weight and the contact.

Yet another type of motion detector is the micro transducer (fig. 99). A transducer is a device that produces an electric current in response to vibration, shock, or motion. This is a very small sensor, generally round, flat, and like a small coin. This sensor has a small interior piezoelectric crystal that is affected by vibration, very loud noises, or the car being hit by something.

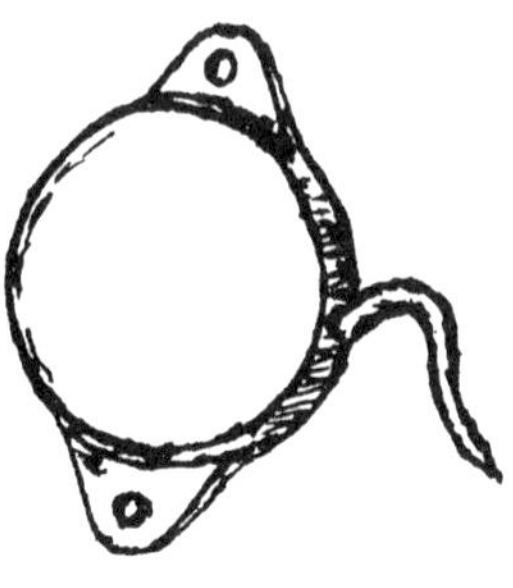

Figure 99. Micro transducer.

A piezoelectric crystal consists of a crystalline material that will develop a voltage when subjected to mechanical stress or severe vibration and, consequently, produce an electric current that triggers the alarm.

Micro transducers are often located inside the door frame near the switch for the interior light, in the same position where you would expect to find a pin-switch. Micro transducers, however, can also be found mounted on the center door posts of four-door vehicles. In either case, they are glued in place with epoxy.

Self-contained vibration sensors are also sometimes used. Such a sensor simply consists of a keypad fitted to the dashboard. The alarm, activated by entering a personal code, works like any other motion detector.

Another type of car alarm sensor is the voltage drop sensor, also known as the current-drain sensor. In this device, a sensor is wired into the electrical circuit of the car. Whenever the sensor detects a drop in the voltage, for instance when the dome light comes on as the door is opened or when the key is put in the ignition and the dashboard lights up, the device will trigger the alarm.

This type of alarm can be disconnected, accidentally or not, by the dome light malfunctioning for some reason. It is therefore not very reliable.

More advanced types of alarms are also used as car alarms. Among these are ultrasonic detectors (see Chapter 7). The most common location for such a sensor is usually on the shelf behind the back seat.

A siren is the best choice for a warning device and is in common use nowadays Various types are available, but they are often mounted under the hood, although well away from sources of extreme heat. The siren is generally mounted slightly downward, as this will prevent excessive accumulation of dirt and moisture. The siren will almost always be connected only to the control unit. Once again, this connection is the vulnerable link in the alarm system. Often the siren will be supplemented by other features, such as a light-flashing facility. Such a device can easily be wired into any of the numerous types of alarm systems.

Finally, remote sensor alarms can be fitted to roof racks and trailers. The really security-conscious can combine all of these options in the same alarm system.

Other specialized features of car alarms include the so-called "passive" alarm system, which automatically arms itself once the owner has locked the car. This alarm is usually armed and disarmed by means of a concealed switch inside the car. Therefore, a certain delay is imposed before the alarm is sounded so that the owner will have enough time to open the vehicle and disconnect the system. Other alarm systems are manually armed and disarmed with an external security key switch, usually located at the rear of the vehicle.

A variant of this system relies instead on an infrared transmitter kept on the owner's key ring or on any other hand-held device. After stepping out of the car, the owner simply aims the device toward the receiver mounted inside the car and presses a button. The infrared beam will activate the system. This will lock all doors and set the alarm. In some versions of this alarm system, the car will even flash its headlights to indicate that the message was received and understood. The alarm can be switched off and the doors unlocked in the same way. Another variant of the same idea is the key ring containing a small radio transmitter that emits a radio signal. Locks of this type are generally called remote control locks.

Another interesting option on many control units is a built-in radio transceiver (a receiver with transmitter). This device will allow the alarm system to be armed and disarmed, or even tripped, by a miniature radio transmitter built into a small pager unit. This circuit will alert the owner to the fact that his car is being tampered with.

Remote paging car alarm systems of this type are especially popular in areas where there is little likelihood of somebody else noticing the alarm if it is sounded. The owner, who carries a remote paging receiver to alert him whenever the alarm is sounded, is then free to investigate himself, call the police, or both.

These systems are sometimes used as silent alarms, without an ordinary siren, especially if the owner hopes to catch the intruder in the act. Some alarms of this type allow the user to choose between siren and silent alarm operation. Those remote paging systems that can arm or disarm the alarm system from a distance almost invariably include a panic button as well with which the owner can activate the car's siren if threatened or surprised by intruders. This option is popular among people who fear walking alone through empty parking lots and professional truck drivers who might have to stay around their vehicles for long periods and even sleep in them frequently.

The remote paging system consists of two parts. The combined transmitter and control unit is mounted in the car, powered by the vehicle's electrical system, while the battery-powered remote paging receiver is small enough to keep in one's pocket. Although these two devices work on a radio channel that is on a frequency seldom used by most radios, they will also be safeguarded by a security code that is always sent when the transmitter is triggered. This will prevent accidental triggering and is also supposed to prevent an intruder from using his own transmitter to disarm the alarm system. The latter is not valid, of course, as an intruder can easily determine and imitate this code if he has access to specialized equipment. (Such equipment is costly, however.)

When the transmitter is triggered, it will send out a signal containing the security code. The paging unit, when detecting a signal on its preset frequency with the proper code, will start beeping. It generally also flashes an LED. The security code is selected in advance by setting a group of internal dip switches. These switches can naturally be found in both devices.

The transmitter will use the vehicle's standard antenna, including the type that automatically rises when the radio is switched on. The intruder can therefore prevent the antenna from rising, which will decrease the range of the transmission severely. The vehicle might of course also have a wire-type antenna built into the windshield, but then the effective range of the transmitter will be significantly reduced in any case. Some vehicle owners attempt to extend the range of the transmitter by installing a separate antenna. Remember, though, that the transmission in this alarm system is the same as the wiring in ordinary alarm systems and should therefore be cut by an intruder. If the alarm call transmission can be prevented, the owner will not be alerted.

Generally, though, the transmission will be sent. The range is then completely dependent on the terrain and the characteristics of the surrounding area. The range might be several kilometers in open country or be reduced to a few hundred meters or less in a city or an underground parking lot. The strength of radio transmissions will also often be reduced considerably by the metal used in most high-rise buildings. Most alarm systems manufacturers claim an "average" range of 3 kilometers, but this is quite an exaggeration. Between 200 and 500 meters is a more typical range within a city. In fact, the transmission can be eliminated by enclosing the transmitter in a metal box. Were it not for the antenna and, to a lesser degree, the windows, the car is, in effect, such a box.

Many car alarms are modified for use in other vehicles and locations, such as boats, trailers, campers, etc. The remote paging systems are especially popular, as they have a great range (more so if connected to a citizen's band base station antenna) and are therefore sometimes used to guard scattered buildings or stores on farms and construction sites.

The car alarm is otherwise armed and disarmed using several different methods, depending on whether a delay is built into the system or not. If so, the delay might be as short as twelve seconds or as long as forty seconds or more. Arming can, for instance, be done by turning the ignition key switch to the ON or ACC (accessory) position for a few seconds and then switching it to OFF. This will arm the alarm system after an exit delay period. It is then disarmed by simply entering the car and switching the ignition key switch to ON before the entry delay period is up.

Another method of arming and disarming is by a switch on the control unit. The easiest, of course, is when the process is performed automatically whenever the ignition is switched on or off. In many cases, the control unit will inform the user of the fact that it is armed or disarmed by either producing a beep or lighting an LED. Sometimes a combination of the two is used and sometimes neither is used.

A valet switch might also be present. This is a switch

that allows the owner to bypass the alarm system when he expects to be away from his car but knows it will be attended or guarded by valet parking or servicing, for instance. The valet switch can only be activated while the engine is running and is therefore of little use to an intruder.

Many car alarms are powered by the car's own battery, but it is also quite possible to add a second power unit as a backup. This is mainly done in commercial vehicles, as on many of these it is easy to gain access to the battery terminals and then disconnect the ordinary alarm. Remember that it might be possible to disconnect the wiring from underneath, even if the battery is locked away under the hood. Having a backup power facility is therefore always prudent.

Almost all car alarm systems are designed to be powered by 12-volt DC current. Furthermore, the wiring is often easy to identify, as a color-coding scheme is common in many countries. For instance, red wiring is used to connect the system to the power source, the car battery. Other colors might be used to identify the components of other systems. Remember, though, that this often varies in different countries—and even in alarm systems manufactured by different companies in the same country.

A professional car alarm system will also have closed circuit wiring. This means that the alarm will sound even if the wires to the sensors are cut. However, the system still will not work if the wire to the warning device is cut instead.

Sometimes the purpose of the entry operation is not only to break into a vehicle but also to move the vehicle in question. This leads to several other considerations, apart from the lock and alarm system.

If the car must be moved, remember that the vehicle might have been immobilized in some way. There are several ways of immobilizing a car. Both electrical and mechanical methods of immobilization can be used, and they should be prepared for.

The electrical means of immobilization include fitting one or more devices, such as an ignition cutout device, to the car. This can be a part of the vehicular alarm system and is then either linked to the car's horn or to any other siren.

The device will sound a warning at the same time as it is blocking the ignition circuit, automatically immobilizing the car.

A manual ignition cutout device is also useful. An ignition disabler switch can be hidden under the dashboard, for instance, and will interrupt the ignition feed wire. Note that the car can still be hot-wired and started easily if the interruption is made between the battery and the coil. If the interruption is made between the coil and the distributor, however, this is not generally possible. The switch might require the use of a key to open.

A passively armed cutout device will render the ignition dead as soon as it is switched off. When the driver wants to start the car again he must deactivate the cutout by depressing a button or switch while he starts the engine.

Yet another method is to use a removable circuit card. Usually this card is put into a socket mounted on the dashboard. When the card is removed, however, vital electrical circuits will be broken, preventing the car from starting. Another method of electrical immobilization is to fit a switch that interrupts the feed wire to the electric fuel pump.

A multiple cutout device can also be incorporated into the central control unit. The unit is bolted to the bulkhead under the hood and disrupts several electrical circuits at the same time. Finally, the vehicle can be immobilized by simply swapping or removing a couple of the spark plug leads.

Mechanical immobilization methods include the use of an engine immobilizer switch, the previously mentioned hidden switch in the ignition circuit, or an internal locking device that is, for instance, fitted over the handbrake and locked around the gear lever. Locks that are simpler, but also fairly reliable, include combination locks attached to the handbrake, engaged or disengaged by means of a three-digit combination. Such locks slide over the top of the handbrake lever, locking it in the "on" position. Electronic locks that prevent the engine from starting until the driver has entered the correct code on a keypad fixed to the dashboard can also be procured.

Additional steering wheel locks are also commercially

Figure 100. Steering wheel hook lock.

available, although they often are of the hook type, hooking on or over the steering wheel and brake, accelerator, clutch pedal (fig. 100), or floor-mounted gear stick. If the device is hooked over the clutch pedal, an intruder can easily get rid of this device in an emergency by simply stamping down hard on the clutch. As the steering wheel bends quite easily, the lock will come off. If the brake pedal is used instead of the clutch pedal, however, this procedure is sometimes more difficult to perform. Of course, the quality of the locking device will also affect the outcome.

Additional locks can be fitted around the steering column like an armored collar. The lock key will then replace the vehicle ignition key and control the electrical operations. Although good in principle, this can easily be demolished with the use of heavy-duty tools.

Some suspicious individuals remove the rotor arm from the distributor in order to protect their cars from theft. The distributor cap is easy and quick to snap off in order to remove the T-shaped rotor arm sitting in the middle of the distributor. It is small enough to put in a handbag or pocket. Others remove it and then lock it in the trunk.

A final, and very definite, method of mechanical immobilization is to use a wheel clamp, also known as a Denver boot. This will effectively prevent the car from being moved. Most wheel clamps also prevent the tire and wheel from being removed. These can be found on cars that are left unattended for considerable periods of time or have been secured by the police.

Vans, trailers, and motorcycles present still other problems, especially if they must be moved from the location in which they are found.

Vans and trailers are sometimes safeguarded from towing by locking a hitch lock, or tow ball, into the ball socket of the vehicle-towing hitch. Such a device can only be removed by using the correct key bar. The lock must be picked or else broken.

Motorcycles and bicycles are generally secured with chain locks. These are generally easy to cut through with bolt-cutters, especially if they are not of made of hardened steel. For this reason, specially designed padlocks with hardened, elongated shackles are often used instead. The steering lock of a motorcycle is usually of simple construction and can often be broken by a fierce wrench of the handlebars. Although motorcycles can be fitted with electrical immobilization devices and alarm systems, this is very uncommon.

It is more common to immobilize the motorcycle by removing the battery ground strap or the line fuse in the main lead near the battery terminals. A concealed cutout switch that breaks any of the low-tension wires to the coil can also be fitted. As long as the correct equipment is available, neither of these methods will present any problems to the operative.

Other measures used to protect a motorcycle for long periods of time include such devious alterations as fitting unserviceable but visibly complete spark plugs, draining the float chambers and removing or blocking the fuel supply line, or putting the bike in first gear and then removing the gear and clutch levers. Once again, should it really be necessary to move the motorcycle, this is only a matter of having the correct spares available.

Shoplifter Detection Systems

Alarm systems designed to detect and scare away shoplifters have been in widespread use for a long time.

Several types exist, and all of them are still very common.

These types of detection systems rely upon fastening some kind of indicator or tag to each object to be protected. In an electromagnetic system, the tag might be a magnetic tape, for instance, while in a radio frequency system, it will be a special coil on a circuit card. In either case, the alarm will sound if the tag is passed near a specially designed detector.

The radio frequency system is easiest to use, as the electromagnetic detectors are very clumsy. However, the radio tags are always active, which might cause complications. The radio frequency system will also sometimes trigger a false alarm if exposed to portable radios. Furthermore, the tag can be rendered inoperative by being hidden in a metallic cover. This will nullify the signal and prevent the alarm from sounding. For these reasons, neither of these systems is very popular.

The British company Securitag International is currently the major manufacturer of shoplifter detection systems. Its products are very popular, as they rely on a completely different technique.

The Securitag system also consists of tags that are affixed to the goods to be protected; however, the actual functioning of the tag is different. The detector posts continuously send out a low-frequency signal that will trigger any tag that is brought within range. The activated tag will then respond by transmitting another signal. This signal will trigger the alarm.

Whenever a tag is brought out through the door, it will transmit a

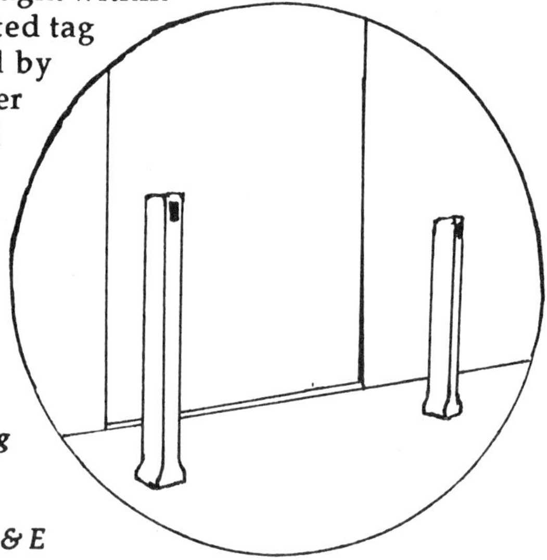

Figure 101. The Securitag system.

signal that is received by the detector posts positioned either near or on either side of the door (fig. 101). Every detector post has a range of about 90 centimeters. Alternatively, a detector loop can be installed around the door, in which case no posts are required.

Because of the design of this system, there is no chance of a false alarm being caused by portable radios or metal objects.

The tags are naturally extremely difficult to remove without the special equipment available from Securitag. However, as this equipment is the same in all Securitag units, an intruder can generally acquire it easily should it ever become necessary.

As a curiosity, it should be mentioned that this idea was originally introduced by the KGB for use in electronic surveillance operations.

FIRE ALARM SYSTEMS

Fire alarm systems are sometimes combined with intruder alarm systems, but the two are more often separate. The warning sound used in fire alarm systems is usually distinctly different from that of intruder alarm systems. This is generally true whether the systems are connected or not, except in the simplest alarm systems.

Whether the fire alarm is incorporated in another alarm system or completely self-contained, it will more often than not rely upon either one or both of two radically different types of sensors. These are the smoke detector and the thermal detector, both of which are very common. Yet another type of sensor, the differential detector, is also used sometimes.

The smoke detector (fig. 102) is placed on the ceiling or high up on a wall, as smoke always rises. Several smoke detectors, either self-contained or alternatively connected in parallel, can often be seen in bedrooms or at least in a common hallway.

There are two different kinds of

Figure 102. Smoke detector.

smoke detectors, the ionization type and the photodiode (or photoelectric) type. They differ in the way they detect the smoke that will trigger the alarm. The ionization type is most popular.

An ionization smoke alarm has a small internal chamber containing a very small amount of radioactive material. The air in the small chamber is ionized and therefore able to conduct an electrical current. Since the radioactive material is located between two electrically charged electrodes, the ionized air will conduct the electrical current between them. If smoke particles (excess carbon particles in the air) enter the chamber, they will increase the resistance of the ionized air. This will naturally decrease the flow of current between the two electrodes. The alarm will sound when the resistance has increased to, and the current flow drops below, a certain preset point. This happens whenever there are too many carbon particles in the air.

The photodiode smoke detector relies upon a beam of light that is projected across a sensing chamber onto a photoelectric cell. When smoke particles enter this sensing area, the light beam is disturbed and the level of light reaching the photoelectric cell is reduced. The alarm is triggered when the level of light reaching the photoelectric cell drops too much.

The reason ionization alarms are the most popular is that they will respond slightly faster to a rapidly spreading fire. Such a fire always produces many smoke particles. Both types are reliable, however. They will even respond to the tiny smoke particles produced by a fire before actual smoke can be seen.

Sometimes a high level of dust in the air will produce a false alarm when it accumulates inside the sensor. Dust in the sensing chamber might also reduce the sensitivity of the alarm. Other sources of false alarms might include small insects or high humidity.

Self-contained smoke detectors have integral batteries, usually of the carbon-zinc or alkaline types. Alkaline batteries are the most reliable, as they last longer. Despite this, they should be checked at least once a week for safety reasons.

Although a reset button is included in the alarm, the continued presence of smoke will simply result in the sounding of the alarm again. Therefore, it cannot simply be turned off as long as the conditions that triggered it remain.

Smoke detectors can generally be used in any room except the kitchen. Thermal, or heat-sensitive, detectors are used there instead.

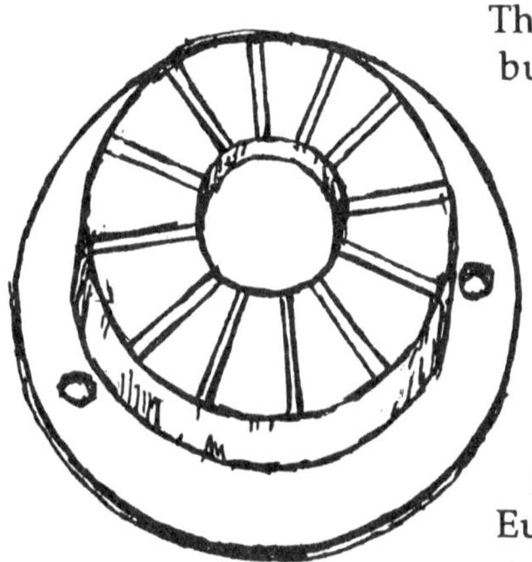

Figure 103. Thermal detector.

Thermal detectors (fig. 103) use built-in pyroelectric sensors able to sense when the temperature in the room rises above a dangerous level. The exact temperature level that is deemed critical depends on the construction of the sensor.

Typically, though, 50°C is the most common in Europe, while 135°F sensors are used in the United States, except in kitchens and furnace areas where a higher temperature is normal. In these areas, 80°C or 190°F sensors are used instead. The sensor contains a material that will melt at the chosen threshold temperature. This will produce electric contact and trigger the alarm.

Thermal sensors can also be wired in parallel. Except in kitchens, they are used mainly in areas where fires that produce more heat than smoke—certain types of electrical and chemical fires, for instance—are likely to occur. Yet another type is the differential detector. This is an advanced thermal detector for use in environments that are unsuitable for ordinary detectors, such as certain kitchens and garages. This detector works by adapting itself to the temperature in the area. It will trigger the alarm when the temperature rises more than 5 degrees per minute.

For a professional intruder, the main advantage of a fire alarm system that is part of an intruder alarm system is the fact that the fire alarm will override the actual intruder alarm. This is very useful, as the deliberate activation of the fire alarm might provide an easy diversion. It is not really necessary to start a fire in order to activate this type of system. Simply blowing smoke into a smoke detector will trigger the alarm. Afterward, there will be no explanation as to why the fire alarm sounded at that particular time. The operative might even go to the trouble of blowing some dust into the detector in order to make it appear as if the alarm malfunctioned.

The fire alarm is easy to find if present. As smoke will to some extent be impeded by doors, most fire departments recommend the installation of several sensors, optimally one in every room. There will be at least one in each hallway, however, and probably one in the kitchen or in other areas where a fire is more likely to break out, such as near heating units and fuse boxes. They are almost invariably mounted on the ceiling, generally at the center of the area they are protecting.

CHAPTER 9

Entry Tricks

This final chapter will be devoted to several of those tricks of the trade that one day might be useful to the field operative who is involved in entry operations. These tricks include ways of finding plans of the target's alarm system, as well as recognizing traps and faked obstacles to the operation. Common hiding places of keys and combination codes also fall under this heading. All these methods help the operative avoid capture when on an entry mission. Every additional advantage increases the security of the operation.

The easiest way to enter the residence of an unsuspecting owner is to use the real keys. If the owner leaves the keys outside his house when he goes away, this is quite easy. The operative should therefore check the seven most common hiding places for keys:

1. under the doormat
2. under a flowerpot
3. in a flowerpot
4. under a stone alongside the path or near the front door
5. stuck or hanging under the window ledge
6. hanging inside the door on a piece of string that can be pulled through the letter box in the front door
7. just inside an unlocked garage or storage shed

If the operative can gain entrance to the house for a few moments, he might be able to check whether the key is left in the lock on the inside or in an obvious cupboard or hook near the front door. If so, he might get an unsupervised moment in which to make a pattern of the key in a wax box. Then he can make a copy at his leisure.

Sometimes supposedly locked doors are nevertheless found unlocked or even opened. The reason is usually very simple—there is work being done on the premises which is facilitated in this way, for instance, or a visitor or delayed employee is expected to arrive shortly. Some people regularly leave the door unlocked, especially in offices, so that they do not need to use the key when they return. But the human factor is not always the reason. The lock may be installed incorrectly, so that the door will not lock even when closed, or the door check does not work properly.

For this reason, many companies employ a guard or a caretaker to regularly check that all doors are locked and remain so. If the guard follows a fixed routine, however, you can plan the operation so that he is safely out of the way. Otherwise it is a good precaution to position a colleague to watch out for the guard so that you know when he is making the rounds.

The points of entry include, but are not limited to, doors, windows, and any other openings that provide access to the interior of the target building. As was mentioned in Chapter 6, it is also sometimes desirable to enter through a roof or ceiling. Most houses protected by an alarm system have a sensor on each outside door or window that is accessible from the ground and large enough to crawl through. Windows on the second floor are generally not protected, however, unless they are easily accessible from a flat garage roof or a tree. If they are accessible, but not easily so, and at the same time cannot be seen from the street or from a neighboring house, they might be protected by alarm sensors. Try to pinpoint the sensors before the actual entry attempt. There might be an opening that has been overlooked.

As every burglar knows, whether he is a criminal or employed by his government, there are certain indications of whether or not a house is empty. One of these, maybe the most important, is the absence or presence of light.

Many houses today are equipped with outside lights that are switched on at dusk every night by a light-sensitive photoelectric cell controller. Such a device measures the intensity of natural light. It will be triggered by the natural light levels and switch on the light, when the light falls below a certain, predetermined level. The lights will be switched off automatically when daylight returns. Other buildings rely on programmable automatic timer switches that regularly turn on the light at a predetermined time.

Some of these timer switches are of the twenty-four-hour variety that will turn the light on and off at the same time every day. Others are the seven-day type that can be set to turn the light on and off at different times each day of the week. The switching pattern will be repeated only after a week has passed. Sometimes the users of these devices even realize that they must vary the setting of the timer with the changing of the seasons. There are devices available for this very purpose. They can be programmed to vary the times the light, or lights, will be turned on each day, according to the seasonal changes in the day's length. These devices are called solar dial timer switches.

Such devices are often both powered by and transmit their signals over the electrical wiring in the building, although remotely controlled units utilizing radio waves are also available. The lighting may even be connected to the output from the alarm system, so that all or at least most lamps in the building will turn on or start to flash whenever the alarm sounds. Other appliances, such as stereo systems, can also be activated in this way. Virtually nothing is impossible, but the sobering fact remains that this will not prevent a professional intruder.

Finally, passive infrared detectors might be used for turning on the light as soon as somebody approaches. Such a detector is easily wired to a light source, and the light will remain on as long as there is somebody in the vicinity emitting body heat or for a set period of time after the source of warmth has gone. Such an infrared floodlight system includes a built-in photoelectric cell that prevents it from triggering during daytime. Sound-triggered detectors are also available for the same purpose.

Another security device is an electric curtain controller, which will open and close corded curtain sets. This device is generally connected to a timer for automatic control.

A radio, preferably tuned to a "talk" station, is often left on or turned on with a timer switch as well. Finally, a telephone answering machine programmed with the message that the owner "cannot get to the phone right now" (not "I'm out") is also frequently used to leave some doubt as to whether anybody is at home or not.

However, these means of scaring away ordinary burglars are not very effective. It is generally easy to check whether anybody is at home or not, by making a personal call pretending to be a salesman for instance. Furthermore, some people leave the light on only in the hallways rather than in the living rooms. This is a definite indication that the house is empty, as nobody actually lives in the hall. Leaving the light on in a downstairs room with the curtains closed it is slightly more clever, but not enough so to deter a professional intruder. Most individuals who use this ruse to protect themselves are instead advised to leave the light on in an upstairs bedroom, where an intruder cannot peer in to check whether it is occupied or not.

It should also be mentioned that cutting the electrical power and/or the telephone lines will effectively determine whether anybody is at home or not, as the owner is certain to reveal himself if he really is at home. Furthermore, it will render useless most of the previously mentioned protective light systems, etc.

Therefore, some people instead use a recording of a barking dog, which is activated whenever a doorbell is depressed. There is even a self-contained alarm system available that will imitate the sound of a dog whenever somebody is approaching the protected area and is detected electronically by a passive infrared detector. These various dog imitation systems can always be identified by the fact that you will only hear the actual barking and no other noise such as the dog running around or jumping at the door.

On the other hand, real dogs are a very serious problem, especially if they bark at strangers and generally appear unfriendly. Guard dogs are not always dangerous, as they are mainly kept as a psychological deterrent to frighten intruders away rather than actually attacking them. Terriers, for instance, make excellent guard dogs for this reason. But whether the dog is really dangerous or not, its bark will call unwanted attention to the scene, so dogs should always be avoided, or at least silenced in some way, if at all possible.

Certain large dogs, such as the Great Dane, the mastiff, the Alsatian or German shepherd, the Doberman, and the rottweiler, are sometimes trained not only to set off the alarm, but also to defend the home against attack. Really dangerous patrol dogs are not usually kept as domestic watchdogs, however, as they are simply too dangerous unless in the care of a professional dog handler. Here, too, it is important to avoid the dog by all possible means.

As a curiosity, it might be noted that in rural areas, geese present a very real obstacle. Geese are highly territorial, and not only do they honk furiously, they also often chase and peck any intruder who does not retreat quickly enough.

Many apartment complexes employ audio entry systems. This is a means of access control in which a speaker panel is located outside the premises near the front door. The speaker is linked to a telephone handset or microphone device that permits two-way speech. The device is frequently used in conjunction with a digital code lock entry system. Each flat in the building is then equipped with a telephone handset as well as a remote control that releases the electrically operated door lock.

The audio entry system might also include an audio-visual component. It will then function as the standard audio entry system but with the addition of closed circuit television (CCTV).

It is, however, easy to circumvent an audio entry system. Just call up anybody in the building and either say that you have lost the code but a "friend" (who you know is a resident in the building and whose name you will mention) asked you to go inside and take care of something in his apartment. If this does not work, call up another one and say that you will have to enter to deliver something (not the mail, though, as the postman will have the correct code and does not need to ask for it). If all else fails, you can always say you want to come inside to deliver free samples of something you believe is desirable to all of the tenants in the complex. Of course, it might also suffice to wait for somebody to arrive and then enter with him or her.

There are also certain popular ways to gain entrance to a private home, during the preliminary reconnaissance of the premises, for instance. Such methods are frequently used by con men. The operative dresses up as an official, a salesman, or in any other suitable disguise. Then he approaches the target.

He might claim to be from the electricity, gas, or water board, for instance. The "electricity board representative" can ask the owner of the house to stand by the fuse box and turn the power on and off while he goes around the house to "check the circuit." The "gas board representative" might ask the owner to watch the meter outside while he "checks the appliances." The "water board representative" might ask the owner to turn the taps in the kitchen while he "checks the flow" in the other parts of the house, such as the bathrooms.

All of these disguises do carry the risk of the owner of the apartment asking for an official identity card, first of all, and then locking himself in his apartment while calling the local office to ask them about you. You can sometimes avoid this by having the telephone number of a colleague printed on your (faked) identity card. However, this sometimes backfires, too, as the real government office will be listed in the telephone directory.

In the early 1980s, a foreign operative who was working under embassy cover in the former Soviet Union planned to enter the dacha, or country villa, of a local party official. The plan was to enter in order to plant electronic bugs in the building. The party official himself was not very important, but his brother was a high-ranking member of the KGB. It was assumed that one could pick up interesting gossip on the occasions when the KGB brother visited the house.

In order to have a first look around, the operative dressed up as an official. He made a personal call to the family in order to make inquiries about the rating of the house. Such inspections were fairly common there, although less so when it came to influential party officials.

The brother of the KGB officer was unexpectedly cooperative, however, and during their conversation, the operative realized that the official had actually asked for a rating to take place. The reason was that his dacha was not big enough. Now he was overjoyed, as he expected the operative to help him find a bigger one.

What was worse, it soon turned out that the real rating official was expected to arrive later the same day. At this point, the foreign operative hurriedly excused himself and left. The operation was abandoned, and the official's house remained free of electronic bugs.

Despite these and other obvious risks, it is amazing how easily most individuals are taken in by such simple tricks. It is always worth a try, as long as an escape route is prepared in advance.

The operative might also disguise himself as a salesman. Then he might offer a free estimate for carpets, furniture, or anything else that will be so attractive in price that it will enable him to enter the premises. A variant of this theme, in rural areas only, is the itinerant antiques buyer who wants to browse around the house looking for interesting pieces to buy.

In rural areas, where open solid fuel fires are common, the field operative can choose to disguise himself as a chimney sweep. A common trick used to be left alone in the house is to tell the owner to go out into the garden and shout when he sees the brush coming out of the chimney.

It is a great advantage to know in advance of the existence and type of any alarm or locking device in the building. In some countries, notably the United States, many cities and counties require a permit for installing an alarm system. Therefore, the relevant archive might be a good source of information regarding the possible existence of any such means of protection.

The plans for an alarm system will often be kept at the security company that originally designed it. The installation codes and operator codes will also be kept there. If you can gain entrance to this company's office, you can take advantage of this information.

Finally, it must be remembered that the local fire prevention unit will have a say when a major alarm and lock system is designed. This might mean that the fire station will have a copy of the plans, including master keys in certain cases. It might also mean that some exits will be left open, or almost open, on purpose, regardless of the risk of an intruder entering. The safety of the personnel working on the premises is generally a higher priority than the ability to keep intruders out. This is especially important in a building where burglars are not expected to be interested in the merchandise or machinery.

Vehicles are nowadays protected in many different ways. In some cars, for instance, the standard locks will have been replaced by new ones. Still, all too many people frequently lose their car keys. For this reason, some cars have a magnetic box containing a spare key that fits under the car. If you find such a box, entry into the car is very easy, of course. Just open the box and retrieve the key.